Laboratory Manual to Accompany

Environmental Science

Action for a Sustainable Future

W. Merle Alexander

Heidi A. Marcum

Daniel E. Beams

The Benjamin/Cummings Publishing Company, Inc.

Redwood City, California ■ Menlo Park, California
Reading, Massachusetts ■ New York ■ Don Mills, Ontario.
Wokingham, U.K. ■ Amsterdam ■ Bonn ■ Sydney
Singapore ■ Tokyo ■ Madrid ■ San Juan

Associate Editor: Laura Bonazzoli
Production Coordinator: Eleanor Brown
Cover Design: Yvo Riezebos

ISBN 0-8053-4223-0

1 2 3 4 5 6 7 8 9 10--AL--98 97 96 95 94 93

The Benjamin/Cummings Publishing Company, Inc.
390 Bridge Parkway
Redwood City, California 94065

PREFACE

The materials and activities presented in this laboratory manual represent a series of topics that are essential to the study of our environment. While the laboratory manual is not intended to correlate exactly with the text book, it does follow the concepts and ideas that are presented, and should give a strong supplement to the text book and its central theme of achieving a sustainable society.

As the student is finding out, environmental problems are extremely broad in scope and it is easy to see elements of many combined disciplines as we try to understand the nature of the world around us. Such disciplines as Ecology, Geology, Chemistry, Physics and Atmospheric Science, and Biology are frequently involved in addressing many environmental problems. As we study the causes and impacts of such problems, we cannot grasp the complete situation without also including Law, Economics, Ethics, and Politics. If we wish to fully understand what is happening in our local, regional, national and global communities, we must realize that environmental problems are not simple ones, and will not be solved with simple solutions.

In considering laboratory activities, we are talking about something that is an activity that directly involves each student, either individually or as part of a group. The materials in this manual, plus the instructions by the teachers, are primarily the tools for the learning process. The value the student gains from the laboratory exercises depends entirely on the student. What the student learns from these activities is strongly dependent on the student's performance; the more the student becomes involved in the lab activities, the more that student will gain from the lab. Thus, the student is urged to become involved in, and to contribute to, each activity.

No single laboratory exercise can provide the student with the total knowledge involved in the mixture of disciplines we find in a complicated environmental problem. However, the fourteen laboratory experiments in this manual treat the fundamental concepts of these disciplines in such a way that, as you begin to achieve the objectives outlined in each experiment, your comprehension of the relationship between these disciplines is greatly enhanced.

The experiments in this manual are grouped into five units. Each unit builds upon, and adds to, the previous unit. By the time we reach the end of this book, we hope that each student is aware of the need for, and the ways of achieving, a sustainable society. Unit I introduces the basic concepts involved in ecology. Since any sustainable ecosystem must have stable species populations, Unit II studies both non-human and human populations. In Unit III, we examine the impact of human-caused environmental degradation on the ecosystem. Since we have a certain amount of finite resources in our ecosystem, a sustainable society must be very efficient in its use of non-renewable resources. Unit IV gives the student a fundamental understanding of the primary role renewable resources must play in our future. Finally, in Unit V, we take a brief look at the interaction of our species with each other as we live and impact the ecosystem we are a part of.

W. Merle Alexander
Professor and Director, Institute of Environmental Studies
Baylor University, Waco, Texas

Heidi A. Marcum
Lecturer, Institute of Environmental Studies
Baylor University, Waco, Texas

Daniel E. Beams
Research Associate, Institute of Environmental Studies
Baylor University, Waco, Texas

Table of Contents:

Chapter 1: Information and Instructions for Laboratory Work

I. Objectives

After completing this introduction the student will be able to:

1. understand the nature of environmental science laboratory techniques

2. apply the scientific method to experimentation in this course

3. write a scientific laboratory report

4. use exponential numbers and metric conversions for calculating laboratory results.

II. Discussion

Environmental science deals with the interaction of human beings and their environments. It is a young discipline when compared to classical sciences such as biology or chemistry. In this course, you will apply techniques developed in the physical, natural, and social sciences to real-world environmental problems. Many of the methods taught are used throughout the world to analyze environmental problems and will allow you to compare your results to those obtained by professionals in the field.

All students should remember that there are no necessarily "right" answers for the exercises in this lab manual. In a chemistry lab, the student can predict results with confidence and may assume that some experimental error has been made if those results differ from predictions. In this course, however, the outcome of each exercise depends on a multitude of variables that are not yet fully understood. Many environmental factors which cannot be controlled by the experimenter are involved. Results may be predicted only in the most general way and may be perfectly correct even if they differ significantly from those predicted. Even so, we may still use the **scientific method** to help recognize what questions we are asking in each exercise, to make generalizations concerning our results, and to organize our data.

Science simply discovers facts and relationships based on observation. Objects or events you try to explain may be found in natural settings or planned as laboratory experiments. The important factor is that they must be directly observable.

Human **observations** are weak because they are based on imperfect sensory perception and may be colored by previous expectations. Observers often find what they expect to find. Even when the unexpected is observed, the observer may disbelieve his or her own observations. New techniques and equipment can make observation easier and more accurate, thus increasing the amount of available data and possibly proving past observations inaccurate.

Once a person has made a series of observations, he or she must do something with the resulting data. Some sort of general statements or **hypotheses** must be made about the data. One type of generalization summarizes and makes a statement about a set of data. For instance, a person may notice that the sycamore tree in his or her yard begins to lose its leaves in late October. After observing other sycamore trees in neighboring yards losing leaves at the same time, the observer concludes that all sycamore trees begin to lose their leaves in late October. This conclusion is an example of a generalization based on inductive reasoning, or **induction**. Induction involves reasoning logically from the specific to the general, from isolated observations to a general statement. Generalizations can be drawn confidently only if a large number of observations have been made which reduce the distorting effects of individual differences.

The inductive generalization can be used as a hypothesis by expressing it as an "if" statement. This hypothesis can be tested by observing other sycamore trees. As more information is gathered, the observer may find it necessary to change the original hypothesis. For instance, the hypothesis may have to be changed to: All sycamore trees in the Southwest lose their leaves in late October.

A second type of hypothesis is an explanatory hypothesis. This type of hypothesis goes beyond a simple summary, and attempts to determine the cause(s) or reason(s) for certain observations. One explanatory hypothesis might propose that sycamore trees begin to lose their leaves because of a change in the photoperiod. Another may relate leaf loss to changes in plant hormone levels. These hypotheses are easily tested by setting up experiments comparing leaf loss under various photoperiod conditions, or by altering plant hormone levels. Both hypotheses meet the scientific requirement of being testable. The hypotheses may be stated in "if .... then" form, for example:

> If sycamore trees begin to lose their leaves in late October because of changes in plant hormone levels, then a large artificially induced change in hormone levels at other times should also result in leaf loss.

Hypothesis testing involves the use of **deductive** reasoning. A person using deductive logic starts with general observations and makes a specific conclusion. The "if" portion of the statement is the hypothesis, and the "then" portion contains the predictions that are based on the assumptions made in the hypothesis. If after experimentation the predictions prove to be false, then the hypothesis is false. In science, hypotheses can never be absolutely proven. Science does not deal with certainties, but with probabilities.

Many false hypotheses have been accepted because they led to true predictions. To show that such a hypothesis is false, other tests must be developed. One false experimental result does not always mean the hypothesis must be totally discarded. It is more a question of the percentage of results which support the hypothesis for it to have any value. Statistical analysis is used to determine the significance of any deviations seen during prediction testing. The larger the sample size or the greater the number of observations, the more likely the hypothesis can be accepted or rejected confidently.

Both qualitative and quantitative data may be collected from experimentation and observation. Scientists prefer quantitative data. This type of data can be presented directly to the reader without subjective interpretation, which may introduce bias. Quantitative data can also be checked easily by other experimenters. Qualitative data are less easily checked and present fewer opportunities for verification than quantitative relationships.

Data must also be organized in such a way as to give the investigator useful information. If the experiment used control groups, the easiest organization would be to compare data from each group. If similar measurements were made at various locations or under various treatment conditions, the data could be organized by group, location, or type of measurement.

The scientific method has some limitations. Researchers using it cannot study any phenomenon which cannot be observed. The scientific method cannot be used to make moral or value judgments. Science can predict human reactions to a given situation and it can provide information to people so they may make moral and value judgments for themselves.

One of the most important things you will do in this class is write **laboratory reports**. Your report is your tool for expressing what you did, why you did it, and what you learned in the process. Even if your understanding of the procedure, techniques, and results is perfect, and your results error-free, a poorly-written report will not indicate that you really understand what you have done. Writing reports is not difficult if you remember a few guidelines. Normally, scientific reports are divided into the following sections:

Abstract

The abstract should contain a brief summary of purpose, methods, results, and conclusions. It should be no longer than three or four sentences.

Introduction

Write your introduction in such a way that the reader will be interested in reading the rest of your report. The introduction consists of two parts:

1. An introduction to the chapter topic and its importance to the environment and society. Cite at least one reference [e.g., (Chiras 1991, p. 32)].

2. A connection between the introduction and why you performed the exercises (this is your hypothesis). End the

introduction with specific questions you intend to answer while testing your hypothesis.

Materials and Methods

This section should include a description of the study area (when applicable) and a summary of what was done and what equipment was used. The procedure should be presented in chronological order and in past tense. Where applicable, number the procedural steps.

Results

Tear out your data sheets and include them with this section. The results section consists of two parts:

1. Original data obtained from using the procedure.

2. Data derived or calculated from the information obtained during the lab exercise.

Discussion

This is the most important portion of the report. In this section, you will discuss your data and any trends or relationships that appeared in the data. Provide insight as to why such trends may have occurred. In this section you discuss whether your results support your hypothesis. Are your results consistent with what you expect? Why or why not? Are your results above or below accepted environmental quality standards? Why or why not? Answer the question asked in the Introduction section. Any problems encountered during the procedure which may have caused errors should be discussed. Pay close attention to both human error and equipment error.

Whenever necessary, suggest other experiments that should be done or additional data that should be collected to answer your initial questions (from your introduction) more thoroughly.

Conclusion

Conclusions are to be based on data, not opinion, and they should follow from your discussion. All conclusions drawn should relate to the statement of the problem (your hypothesis). Did your results confirm or deny your hypothesis? Were your results above or below accepted standards?

Literature Cited

In the text of your report, cite references using the author's surname, year of publication, and page number [e.g., (Chiras 1991, p. 32)]. In the bibliography, use the following citation format:

Chiras, D. D. Environmental Science, Third Edition. Redwood City, CA: Benjamin/Cummings Publishing Company, Inc.; 1991.

Helpful Hints

When writing your report, the following hints may be helpful:

1. Do not put off writing your lab report until the night before it is due. Write it as soon as possible after completing the exercise.

2. Write in active voice. For example: "I shook the solution until it was thoroughly mixed." rather than "The solution was shaken until it was thoroughly mixed." The pronouns we, he, and she may also be used where applicable; for instance, "As a group, we compiled data from two sources." In some cases, it may be necessary to use the passive voice, but it is not a preferred usage.

3. Write in past tense unless it is ridiculous to do so, or unless past tense makes the meaning unclear. The use of past tense is a convention of technical writing.

4. Put section headings in your report. These make the report easier to grade and will give it a professional appearance.

5. Put your data into graphs or tables whenever possible. Be sure to carefully title all graphs and tables. Refer to these by number in the text of your report.

6. Strive for a professional appearance in your work. Write as though you are being paid to do so. Neatness, clarity, style, and appearance of lab reports are important.

7. Even though you may have worked in groups, write your own lab report.

Before coming to class, you should read and understand the material in the lab manual. Each class will begin with a lecture by your instructor, which includes additional information and an explanation of the procedure. Take careful notes because information presented in the lecture may be necessary to complete the exercise and may be required in your lab write-up.

Take notes during the procedure on anything you observe that could have an impact on the results of the exercise. Record this information on the data sheets. Follow the procedure exactly so that experimental results will be valid. Record your data in a neat and logical format.

At the conclusion of an exercise, the instructor will lead the class in a discussion of what has been discovered. Pay close attention to both the written and oral instructions for writing the report. Since each unit has slightly different requirements, the instructor will explain how to write your report. Do not leave class without a complete understanding of how to write your report. Be sure to address each of the questions (listed in the Question section of each chapter) in your lab write-up.

This course is an introductory-level course which usually counts as a lab-science credit for a variety of majors. Therefore, students in this class come from a wide range of science and non-science backgrounds. Some of you may be familiar with the concepts presented in class, while others are getting their first introduction to the material. Contribute your knowledge in class discussions, listen to your classmates, and ask questions. You may discover new viewpoints and aspects of what you thought were familiar issues. For those of you who do not have a strong science background, do not panic -- you will have just as much opportunity to learn as anyone else. Feel free to participate in class discussions and to ask questions. Never hesitate to ask a question because you think it is silly. Some of the best discussions start from supposedly silly questions.

III. Activity

Environmental science, like other natural and social sciences, often deals with quantitative data. Therefore, you should be familiar with the mathematics behind basic scientific methodology.

Any number can be expressed in exponential form. An **exponential number** consists of a number times a power of ten. This is a convenient way to express very large or very small numbers. For example:

$$1,000 = 10 \times 10 \times 10 = 10^3$$

$$8,000 = 8 \times 10 \times 10 \times 10 = 8 \times 10^3$$

$$4,500 = 45 \times 10 \times 10 = 45 \times 10^2$$

$$= 4.5 \times 10 \times 10 \times 10 = 4.5 \times 10^3$$

$$\frac{1}{10000} = 0.0001 = 1.0 \times 10^{-4}$$

Remember, the product of the number times the appropriate power of ten must remain the same. If the decimal point is moved to the left, the power of ten should be increased by the number of places moved; if the decimal point is moved to the right, the power of ten should be decreased by one for each place moved.

Exponential numbers may be **added** or **subtracted** if they are all expressed as the same power of ten.

$$4.5 \times 10^{-4} + 82.0 \times 10^{-3} + 420.0 \times 10^{-6}$$

$$
\begin{array}{r}
4.5 \times 10^{-4} \\
820.0 \times 10^{-4} \\
4.2 \times 10^{-4} \\
\hline
828.7 \times 10^{-4}
\end{array}
$$

When exponential numbers are **multiplied**, the powers of ten do not have to be the same. The numbers are multiplied in the usual way and the exponents are added to obtain the product of the power of ten.

$$(8 \times 10^9)(4 \times 10^6) = 32 \times 10^{15} = 3.2 \times 10^{16}$$

$$(7 \times 10^{-3})(4 \times 10^{-2}) = 28 \times 10^{-5} = 2.8 \times 10^{-4}$$

Exponents are **divided** by subtracting the power of the denominator from the power of the numerator.

$$\frac{54 \times 10^4}{6 \times 10^2} = 9 \times 10^2$$

$$\frac{64 \times 10^4}{8 \times 10^{-5}} = 8 \times 10^9$$

Exponential numbers can also be **raised to another power** by multiplying the exponents.

$$(3 \times 10^4)^3 = (3)^3 \times (10^4)^3 = 27 \times 10^{12}$$

$$(9 \times 10^4)^{1/2} = (9)^{1/2} \times (10^4)^{1/2} = 3 \times 10^2$$

The **metric system** is used by scientists worldwide. However, the United States has been slow to accept the metric system. There have been several national attempts to shift to the metric system;

however, none have been very successful. For laboratory purposes, you should be familiar with the metric system and with methods used to convert measurements expressed in English units to metric and vice versa.

Conversion factors are used to change a measurement to a larger or smaller unit within the same system, or to change the unit of measurement from one system to the other. Conversion from one metric unit to another is very simple: multiply or divide by the proper multiple of ten. For example: 1 m = 1/1000 km; 1 cm = 1/100 m; 1 cm = 10 mm; etc. Any conversion factor may be written in reciprocal form. For example:

$$\frac{1 \ m}{100 \ cm} \qquad or \qquad \frac{100 \ cm}{1 \ m}$$

When using conversion factors, it is best to write down each step of the conversion. There may be only one step or there may be many, depending on the initial units of measurement, the desired unit, and the conversion factors chosen. There may be more than one correct path available for any given conversion. Answers obtained from the various paths may vary slightly due to rounding of conversion factors. For example, to change kilometers (km) to feet (ft), you can use either of the following two conversion formulas:

$$\frac{35 \ km}{1} \ x \ \frac{(1000 \ m)}{(1 \ km)} \ x \ \frac{(1 \ ft)}{(0.305 \ m)} \ = \ 114754.09 \ ft = 1.15 \ x \ 10^5$$

$$\frac{35 \ km}{1} \ x \ \frac{(0.621 \ mi)}{(1 \ km)} \ x \ \frac{(5280 \ ft)}{(1 \ mi)} \ = \ 114760.8 \ ft \ = 1.15 \ x \ 10^5$$

Conversion factors must be used in the form that will cancel the initial unit and leave the desired unit. In the example above, kilometers and meters cancel out, leaving the final answer expressed in feet. If either conversion factor had been expressed in its reciprocal form, an incorrect answer both numerically and in terms of unit would have been obtained.

$$\frac{35 \text{ km}}{1} \times \frac{(1000 \text{ m})}{(1 \text{ km})} \times \frac{(0.35 \text{ m})}{(1 \text{ ft})} = 10675 \text{ m /ft} \quad \text{INCORRECT}$$

Even if a conversion problem appears difficult or complicated, you should have no problem if you have the necessary conversion factors and write each step of the conversion down. Use the conversion factors in a form which cancels out the initial unit and each intermediate unit until you are left with the desired unit. If you started with a unit of length, you should end up with a unit of length. The same holds true for units of area, volume, and all other dimensional units. For example, if you want to convert 5 square feet to square centimeters you could use the following steps:

$$\frac{5 \text{ ft}^2}{1} \times \frac{(12 \text{ in})^2}{(1 \text{ ft})^2} \times \frac{(2.54 \text{ cm})^2}{(1 \text{ in})^2} = \frac{(144 \text{ in}^2)}{(1 \text{ ft}^2)} \times \frac{(6.452 \text{ cm}^2)}{(1 \text{ in}^2)}$$

$$= 4645.44 \text{ cm}^2$$

Notice that each conversion factor must be squared: the factors are in one-dimensional units of length, and they must be squared to obtain the two-dimensional units of area. Notice also that squaring means multiplying the number by itself, not multiplying it by 2 or simply squaring the unit. The same holds true for cubic units. The number is multiplied by itself three times. For example:

$$12^3 = 12 \times 12 \times 12 = 144 \times 12 = 1728$$

$$\frac{5 \text{ ft}^3}{1} \times \frac{(12 \text{ in})^3}{(1 \text{ ft})^3} \times \frac{(2.54 \text{ cm})^3}{(1 \text{ in})^3} = \frac{(1728 \text{ in}^3)}{(1 \text{ ft}^3)} \times \frac{(16.387 \text{ cm}^3)}{(1 \text{ in}^3)}$$

$$= 141583.68 \text{ cm}^3$$

The following are examples of various conversions which may be used as guides.

5 ft 8 in = _____ m

$$\frac{68 \text{ in}}{1} \times \frac{(1 \text{ m})}{(39.4 \text{ in})} = 1.73 \text{ m}$$

3 gal = _____ L

$$\frac{3 \text{ gal}}{1} \times \frac{(4 \text{ qt})}{(1 \text{ gal})} \times \frac{(1 \text{ L})}{(1.06 \text{ qt})} = 11.32 \text{ L}$$

80 °F = _____. °C

$$\frac{80°F - 32.0°F}{1.80} = 26.67°C$$

IV. Procedure

Materials

data sheet
calculator
conversion chart

Methods

1. Read and thoroughly understand the principles for performing metric conversions.

2. Use the metric conversion chart in the Appendix to complete data sheets 1.1 and 1.2.

3. Be sure to show your work. Round off all answers so that two digits are to the right of the decimal. Use scientific notation for very large or very small numbers.

Data Sheet 1.1

Name _____   Date _____

1. 5 ft   = _____ c m

2. 3.6 ft   = _____ m m

3. 11 ft   = _____ m

4. 5 ft 8 in   = _____ c m

5. 31 mi   = _____ k m

6. 0.67 km   = _____ m

7. 18 mm   = _____ in

8. 85 cm   = _____ in

9. 73 m   = _____ y d s

10. 32.9 km   = _____ f t

11. 9 oz   = _____ g

12. 15.7 lb   = _____ k g

13. 25 kg   = _____ l b

Data Sheet 1.2

Name _____   Date _____

14. 1 qt 7 oz = _____ l

15. 325 gal = _____ l

16. 1226 l = _____ ft^3

17. 16 m^2 = _____ ft^2

18. 36 km^2 = _____ mi^2

19. 53 in^2 = _____ cm^2

20. 442 in^3 = _____ cm^3

21. 7 ft^3 = _____ m^3

22. 117 cm^3 = _____ ft^3

23. 212 °F = _____ °C

24. -46 °C = _____ °F

25. 121 °C = _____ °F

Unit I

Ecology

Ecology is the foundation for every action and reaction that occurs in the environment. Studying organisms and their interactions with the environment will lead us to a greater understanding of ourselves. It is becoming clear that any trespasses we commit against nature we commit against ourselves.

Unit I introduces basic concepts of ecological communities. Chapter Two describes the diversity of community structure and function. Chapter Three discusses the interactions between organisms in a community. Chapter Four analyzes the competition of organisms for shared or limited resources. Finally, Chapter Five addresses the dynamic and constantly changing processes of succession in communities.

Chapter 2: Community Structure

I. Objectives

The student will gather and analyze data on community structure. After completing the work associated with this chapter, the student will be able to:

1. use the quadrat method for studying communities

2. determine the density of species in the community

3. determine the frequency of each species in the community

4. recognize the importance of community structure in the function of an ecosystem.

II. Discussion

The area in which all life exists is called the **biosphere**. The biosphere is divided into enormous regions which are characterized by climatic or geographic variations. Some examples of terrestrial regions, called **biomes**, are tundra, grasslands, deserts, and tropical rain forests. A subdivision of a biome, an **ecosystem**, is an interdependent, complex network of living (**biotic**) and non-living (**abiotic**) components. All components of an ecosystem interact with each other in a myriad of ways.

All living organisms in an ecosystem are defined as the **biological community**. Plants, animals, and microorganisms are part of their community. Abiotic components of the environment, such as precipitation, temperature, and soil, help define that community. The place in the community an organism may live is called its **habitat**. Organisms of the same species that live in the same habitat compose a **population**.

Humans can learn a great deal by studying communities. Because we are a part of nature, we must learn to live within the limits imposed upon us by our natural communities. Many of our activities, such as urban development and modern farming practices, have had unpredicted effects on the structure and function of communities. By studying both natural and disturbed communities, we can learn to minimize undesirable effects.

Communities are incredibly diverse, both in structure and function. It would be impossible to study every organism in a community, but we can study samples, and gain insight into the function of the entire community. Data may not be exact, but are accurate enough to approximate the entire community.

Even though it is important for ecologists to study plant and animal life and related abiotic factors of a community, this study is limited to quantifying vegetation structure. Vegetative studies are easiest because plants are stationary, whereas animals, even insects, are mobile.

There are many ways to quantify data; some of the more important include: counting the number of individual plants, biomass, area occupancy, frequency, dominance, relative importance, density and cover. Plant samples can be measured directly (i.e. by clipping and drying) and indirectly by using statistical methods. Different habitats (i.e. forests, fields, deserts, etc.) will require slightly different sampling strategies. The most frequently used methods are: quadrat, quarter, transect (strip and line), and distance-to-nearest neighbor.

To get a more complete picture, other procedures, such as aerial photography and analysis of important abiotic factors (i.e. erosion), are also included in ecologists' studies. Studying communities involves gathering a great deal of data; however, this exercise will cover only the first procedure usually performed by ecologists: surveying community structure. This study will focus on surveying a grassland-type community.

III. Activity

This activity will quantify vegetation structure of a community; therefore, the quadrat method is the most appropriate to use. The quadrat method uses plots of fixed size (usually one square meter), and shape as sampling units. Plots can be square, round, or rectangular. In gathering data, many small samples are more informative than a few large ones. The total sampling size should be at least 10 percent of the study area to accurately predict community structure. For example, if you decide the study area will be 100 square meters, you need to sample at least ten 1 square meter plots.

The position of plots surveyed should be determined randomly to eliminate bias. Several methods are used to randomly select plots, such as consulting random number tables, throwing a stick over one's shoulder, or, as in this activity, drawing numbered chips out of a bag. Ecologists record field observations using various methods. Some methods require removing the plants from the field, but we will record the plant species and location on a vegetation map of the community.

A primary function of surveying community structure is determining the abundance of species in relation to habitat, time, each other, or different community types. The measures of abundance are density, frequency and dominance. You can then go on to more sophisticated analyses and comparisons, such as the t test and chi-square test. Density is defined as the number of individuals per area sampled. Relative density is the density of a given species in relation to the total density of all species. Frequency is the number of quadrats in which a species occurs divided by the number of quadrats examined. Relative frequency is the frequency of a given species in relation to the total frequency of all species. From relative density we can determine which species is most abundant.

By comparing different quadrats, frequency determines distribution of species in a community. Relative frequency can tell us if a given species is distributed randomly, uniformly, or in clumps. If the relative frequency of a species is between 0-30 percent, the species occurs in clumps. If it is between 31-80 percent, the species is randomly distributed and if it is between 81-100 percent the species is uniformly distributed throughout the community.

$$\text{Density} = \frac{\text{no. of individuals}}{\text{area sampled}}$$

$$\text{Frequency} = \frac{\text{no. of quadrats in which species occurs}}{\text{no. of quadrats examined}}$$

$$\text{Relative density} = \frac{\text{density of a given species}}{\text{total density of all species}} \times 100$$

$$\text{Relative frequency} = \frac{\text{frequency of a given species}}{\text{total frequency of all species}} \times 100$$

_____

IV. Procedure

<u>Materials</u>

   meter stick
   string
   12" wooden stakes
   notebook
   data sheets
   field guide of local flora
   numbered poker chips

<u>Method</u>

1. Determine the location of your sample community. (This procedure describes sampling in a grassland environment. If a grassland is not available in your area, other community types may be used with slight modification of procedure.)

2. Answer the questions on data sheet 2.1.

3. As a class, identify five major species and record them on data sheet 2.2. (Be certain the entire class uses the same letter to represent the same species.)

4. Define the total study area by stepping off the desired number of meters. (i.e. for a 100 sq. meter plot, step off 10 meters to a side and mark each corner with a stake; using this size as an example, a minimum of 10 quadrats will have to be surveyed.)

5. Working in groups of two, locate a quadrat randomly by drawing numbered chips from a bag. There should be as many chips as there are meters to a side. For example, in a 10 x 10 sq. meter

plot, there should be ten chips in the bag. The first chip drawn from the bag will be the x-axis intersect. Replace the chip and draw another; this will be the y-axis intersect. Record x-y intersect on data sheet 2.2.

i.e., chips drawn
7:     x-intercept
3:     y-intercept

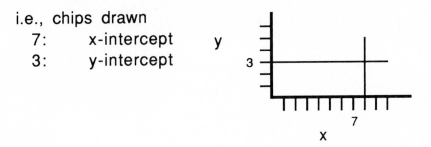

6. Once you locate the x and y intersect, you will lay out a one square meter plot with the x-y intersect being approximately the center. A square quadrat may be laid out as follows. Set a corner stake and lay out a line of one meter for a side. Drive another stake and continue in this fashion until you complete a square. Tie a string around all four stakes.

7. Count all individual plants belonging to the five species listed at the top of data sheet 2.2. Record species type, size, and location on the vegetational community map on data sheet 2.2. Indicate size by drawing a circle proportional to the size of the plant in the quadrat. Record species using the corresponding letter in the above table. While counting, you must make several decisions. To include a plant in the count, at least 50% of it must be in the quadrat. For bushes and grasses you must make a decision as to what is an individual plant (your instructor will help you on this).

8. If time permits, and at least 10 percent of the study area has not been surveyed, survey another quadrat and record results on data sheet 2.3.

9. In the classroom, tally the number of plants in each species by quadrat. Compile results on the chalk board and then record class results on data sheet 2.4, section F. Complete calculations on data sheets 2.5 and 2.6. Answer questions in section V.

Data Sheet 2.1

Name _____   Date _____
Group Number _____

A) Answer the following questions about the environment of the surveyed community:

1) In what biome would you classify the community? _____
   _____
   _____

2) Describe the topography of the community. _____
   _____
   _____

3) Describe local seasonal and yearly climate (temperature, rainfall, humidity, light conditions). _____
   _____
   _____
   _____

4) Describe the soil type (clay, sandy, loamy, etc...), color, texture, organic content, wetness, and cohesiveness. ____
   _____
   _____

5) Describe the animal community (large animals, rodents, and insects). _____
   _____
   _____

Data Sheet 2.2

Name _____

Date _____

Group Number _____

Quadrat _____

Location of Community: _____

Record of Data:

B) List five species in selected community

Species **a**: _____
Species **b**: _____
Species **c**: _____
Species **d**: _____
Species **e**: _____

C) Vegetational Community Map

Data Sheet 2.3

Name _____ Date _____
Group Number _____ Quadrat _____

Location of Community: _____

Record of Data:

D) List five species in selected community

Species **a**: _____
Species **b**: _____
Species **c**: _____
Species **d**: _____
Species **e**: _____

E) Vegetational Community Map

Data Sheet 2.4

Name _____   Date _____
Group Number _____

F) Record the number of plants counted by species in each quadrat surveyed in the community.

Quadrat	Species a	Species b	Species c	Species d	Species e	Total
1						
2						
3						
4						
5						
6						
7						
8						
9						
10						

Data Sheet 2.5

Name _____   Date _____
Group Number _____

G) Calculations:

1) What is the total **density** for all plants in all quadrats?

   a) total number of plants: _____ plants
   b) total number of quadrats: _____ sq. meters
   c) total density = (a/b): _____plants/sq. meter
       (use answer in question 3, column 4)

2) What is the **frequency** for all quadrats sampled?

   a) number of quadrats where plants occur: _____
   b) number of quadrats surveyed: _____
   c) frequency = (a/b): _____
       (use answer in question 4, column 4)

3) What is the **relative density** of each species?

Species	total # of plants (1)	# of quadrats (2)	(1)/(2) (3)	total density (4)	(3)/(4) (5)	X 100 (6)	relative density (7)
a							%
b							%
c							%
d							%
e							%
						Total =	%

28

Data Sheet 2.6

Name _____   Date _____

Group Number _____

4) What is the **relative frequency** of each plant species?

Species	# of quad. species occurs (1)	total # of quadrats (2)	(1)/(2) (3)	total freq. (4)	(3)/(4) (5)	X 100 (6)	relative freq. (7)
a						X 100	%
b						X 100	%
c						X 100	%
d						X 100	%
e						X 100	%

V. Questions

1. How do the environmental factors described on data sheet 2.1 affect structure and distribution of plants species surveyed?

2. According to the vegetational community map which species appeared to be most dominant?

3. Based on the relative density of each species, which plant is the most dominant? Does this correspond with your answer in question two?

4. Based on the relative frequency of each species, which plant occurs most frequently? Is this the same species that was most dominant?

5. What is the difference between density and frequency?

6. According to the relative frequency and relative density of each species, are the plants in the community uniformly distributed randomly distributed, or clumped in groups? (Answer for each species and as a whole community.)

7. What difficulties could you encounter in trying to run quadrat studies on animal populations? In an aquatic environment?

VI. Suggested Readings

Chiras, D. D. Environmental Science: Third Edition. Redwood City, CA: Benjamin/Cummings Publishing Company; 1991.

Cooperrider, A.Y.; Boyd, R.J.; Stuart, H.R. eds. Inventory and Monitoring ofWildlife Habitat. Denver, Co.: U.S. Dept. Inter., Bur. Land Manage. Service Center; 1986.

Darnell, R. M. Organism and Environment: A Manual of Quantitative Ecology. San Francisco, CA: W. H. Freeman and Company; 1971.

Del Giorno, B.J.; Tissair, M.E. Environmental Science Activities: Handbook for Teachers. West Nyack, N.Y.: Parker Publishing Company, Inc.; 1975.

Kaskel, A.; Hummer, P. J. Jr.; Daniel, L. <u>Biology: Laboratory Experiences</u>. Columbus, OH: Charles E. Merrill Publishing Company; 1985.

Lemon, P. C. <u>Field and Laboratory Guide for Ecology</u>. Minneapolis, Minn: Burgess Publishing Company; 1962.

Myers, W.Y.; Shelton, R.L. <u>Survey Methods for Ecosystem Management</u>. New York, NY: John Wiley & Sons; 1980.

Smith, R.L. <u>Ecology and Field Biology</u>. New York, N.Y.: Harper & Row, Publishers; 1966.

Southwood, T. R. E. <u>Ecological Methods: With Particular Reference to the Study of Insect Populations</u>. New York, NY: Chapman and Hall; 1978.

Wratten, S.D.; Fry, G. <u>Field and Laboratory Exercises in Ecology</u>. Scotland: Edward Arnold (Publishers) Limited; 1980.

Chapter 3:
Habitat and Niche

I. Objectives

The student will gather and analyze data from forest habitats concerning niche characteristics. After completing the work associated with this chapter, the student will be able to:

1. determine niche interrelationships of tree species in the field

2. use a field guide to identify local tree species

3. observe niche width and niche overlap between tree species in different habitats

4. recognize the environmental factors to which an organism must adjust for survival.

II. Discussion

Community life is the product of countless interactions between organisms. Many plants and animals occupy the same habitat and must share available resources. The role an organism plays in its community is called its **niche**. A niche is part of the set of relationships of a species to the environment. Variables in a niche include territory, feeding habits, breeding habits, and behavior. Many species can coexist in a community because they occupy different niches.

The **fundamental niche** of a species is the set of all environmental conditions that permit it to exist. The species could exist at every point within which environmental conditions permit. However, even if a species has the potential to exist in a particular habitat, it may not. That portion of the fundamental niche which is occupied is called the **realized niche**. Competition and other species interactions determine what portion of the fundamental niche becomes the realized niche.

One cannot directly observe a niche in the field because it is more of a concept (defined by an infinite number of variables) than a quantifiable measurement. Niche consists of a species' habitat as well as function. The spatial components of a niche (soil moisture, elevation, etc,) define the species habitat. A species function in the habitat is the difference between the fundamental and the realized niche. A niche can be defined for an individual organism (every organism has its own niche), or a niche can be defined for a species. Every organism in a species lives within the limits of its species' niche; however, organisms of the same species do not function in exactly the same manner.

Carrying capacity is the number of organisms a community can support while remaining stable. Carrying capacity can be spoken of both in terms of total number of organisms in a community and the number of organisms of a certain species within its niche in the community. Ecologists often debate the actual carrying capacity of a community because any variation in the number of organisms in a species will change the balance between species in a community. Generally, populations within a community run in cycles that oscillate over a period of years. If a population rises above the carrying capacity, then that community will be permanently altered.

Species **diversity** also affects the stability of a community. A community that supports a large number of species correspondingly has a large number of niches. Stability is simply the ability to resist change. Change is less likely to affect every niche in a diverse community than it is likely to affect a community with only one or two species. A single plant disease may have little effect in a tropical rainforest, but it may completely destroy a cultivated monocrop field.

III. Activity

The niche concept is abstract and impossible to measure exactly. Ecologists have devised methods of calculating values of various indices for niche breadth and niche overlap. **Niche breadth** is defined as the diversity of resources used by a species. It can be used to measure the distribution of species over environments. Some organisms are able to live in almost any environment (i.e., rats, cockroaches), while others are severely restricted in their resource use (i.e., pandas and koalas eat only one type of vegetation). Niche breadth is commonly measured by diversity indices using statistical methods (i.e., Shannon diversity index: H'). However, we will limit our study to observing how several species of plants are proportionally distributed in different habitats.

Niche overlap is the extent to which species share the same resources. The competitive exclusion principle states that no two species can share exactly the same niche; however, species in a community will overlap to some extent. For example, an owl and hawk may share the same food source, but hunt at different times of the day. Or, two plants may share the same soil type, but have different methods of pollination (i.e., wind, insects). There are several statistical measures of niche overlap. However, we will limit our study to observing niche overlap of several species between habitats.

In this activity we will survey tree species in three different habitats. We will record the type and number of trees; however, we will not map their location. We can observe niche breadth and overlap by calculating the mean of the three habitats and then entering all four data sets on a graph. From looking at the chart and the graph, we can observe niche breadth in several ways: 1) species that deviate most from the average have narrow niches; 2) species closest to the average have broad niches; and 3) habitats with the straightest horizontal lines have the greatest number of niches. Different species with approximately the same numbers in two or more habitats have some niche overlap.

IV. Procedure

<u>Materials</u>

    4   three-ft. wooden stakes
    4   colored pencils
         50m. tape measure
         data sheets
         field guide to local tree species

Method

1. Choose three wooded communities with diverse habitats (ie. hillside, river bottom, plateau, different soil types, etc).

2. Go to the first community and describe the environment by answering the appropriate questions on data sheet 3.1. The instructor should help identify the tree species in the community.

3. Measure a 30 meter by 30 meter square plot and mark the corners with the stakes.

4. Divide into groups (depending on the number of tree species in the study plot and the number of people in class). Assign a species of tree to each group. Each group is responsible for identifying every tree of its assigned species within the plot and recording this number on data sheet 3.2, section B. Be sure that all species in the plot are recorded. (Do not count seedlings.)

5. Repeat steps one through four for the next two habitats and record all data.

6. Share results in the classroom and record them on data sheet 3.2, section C. Calculate the average for each species.

7. Graph species numbers on data sheet 3.3 (use different colors to plot habitats A, B, C, and the species average).

Data Sheet 3.1

Name _____   Date _____
Group Number _____

A) Answer the following questions about the three habitats
   surveyed.

   1) Describe your impression of the biotic environment (density,
      ground cover, maturity, etc.).
      Habitat A) _____

      _____

      Habitat B) _____

      _____

      Habitat C) _____

      _____

   2) Describe the topography of the habitat (slope, location, etc).
      Habitat A) _____

      _____

      Habitat B) _____

      _____

      Habitat C) _____

      _____

   3) Describe the soil type, color, texture, organic content,
      wetness, and cohesiveness.
      Habitat A) _____

      _____

      Habitat B) _____

      _____

      Habitat C) _____

      _____

   4) Record other details differentiating the habitats.
      Habitat A) _____

      _____

      Habitat B) _____

      _____

      Habitat C) _____

      _____

Data Sheet 3.2

Name _____   Date _____

Group Number _____

B) Tally the number of trees in your groups' assigned species.

  1) Habitat A -- species _____   tally _____

  2) Habitat B -- species _____   tally_____

  3) Habitat C -- species _____   tally_____

C) Tree Species Data Chart -- record the number of trees of each species per habitat.

Species' common name	Habitat A	Habitat B	Habitat C	Species Total (T)	Average (T/3)
Total					

37

Data Sheet 3.3

Name _____   Date _____

Group Number _____

D) Graph the number of trees in each species for each habitat and the species averages.

Key:
Habitat A -
Habitat B -
Habitat C -
Habitat D -

Number

Species

V. Questions

1. How do the environmental factors in the different habitats described on data sheet 3.1 affect the number and types of tree species of trees in each habitat? Density of trees in each habitat?

2. Based on data sheets 3.2 and 3.3, which tree species have the narrowest niches? Broadest niches? How did you come to this conclusion?

3. Which of the three habitats has the greatest number of niches? How did you come to this conclusion?

4. Which tree species show some niche overlap? Explain how you came to this conclusion.

5. How could you further study the niche breadth and overlap of tree species in your region?

6. From your observations, describe and give an example of how a species adjusts to better survive in its environment.

VI. Suggested Readings

Chiras, D. D. Environmental Science: Third Edition. Redwood City, CA: Benjamin/Cummings Publishing Company; 1991.

Cooperrider, A.Y.; Boyd, R.J.; Stuart, H.R. eds. Inventory and Monitoring ofWildlife Habitat. Denver, Co.: U.S. Dept. Inter., Bur. Land Manage. Service Center; 1986.

Cox, G.W. Laboratory Manual of General Ecology. Sixth edition. Wm. C. Brown, Publishers; 1990.

Darnell, R. M. Organism and Environment: A Manual of Quantitative Ecology. San Francisco, CA: W. H. Freeman and Company; 1971.

Margalef, R. Perspectives in Ecological Theory. Chicago, Ill.: The University of Chicago Press, Ltd; 1968.

40

Pielou, E.C. <u>Ecological Diversity</u>. New York, N.Y.: John Wiley & Sons; 1975.

Whittaker, R.H. and Levin, S.A. <u>Niche: Theory and Application</u>. Stroudsburg, Pa.: Dowden, Hutchinson & Ross, Inc.; 1975.

Chapter 4: Ecological Competition

I. Objectives

This chapter will address ecological implications of competition in a community. After completing the work the student will:

1. use the distance-to-nearest neighbor method to survey competition in a woodland community

2. examine basic properties of interspecific and intraspecific competition

3. demonstrate basic features of competition as an ecological process.

II. Discussion

All organisms compete for food, space, water, minerals, and sunlight. **Competition** is defined as a struggle for shared or limited resources. Since Darwin, ecologists have recognized that competition is an important ecological concept. It plays a major role in community structure, niche breadth and overlap, community succession, and evolution.

Two species compete with each other when they seek the same resources which are in short supply. When organisms compete, they interact with each other in ways that affect their growth, reproduction and survival. Immediate effects of competition include stunted growth and suppressed population levels. The competitive exclusion principle (Chapter 3) holds that competing species will either adjust and coexist, or one will be driven to extinction. The final outcome will depend on the degree of competition. If the degree of competitive impact is low, their niches are different enough that they can coexist. If competition is intense, only the species best adapted to that environment will survive. When environmental conditions change, another species that is better suited may drive out the original "winner".The best way to survive competition is to avoid competition. When confronted with competition, the simplest response is for a species to move. Obviously, individual plants do not move, but as a whole population, plant species can relocate. Another strategy a species may adopt is to secret a substance that is harmful to other species, or even to other members of its own species. For example, species can change their feeding habits, times of feeding, and/or places of feeding through ecological adaptations. Over many generations, competitors may change genetically, favoring the gene that enables the species to exploit a part of the habitat that its competitor does not occupy.

Ecologists debate whether competition is more intense within a species or between different species. Within a community, organisms compete with other organisms of its own species, this is termed **intraspecific** competition. These same organisms must also compete for resources with members of other species, this type of competition is termed **interspecific** competition. Interspecific competition confines species to their realized niches, instead of to their fundamental niches.

In competition, stronger species tend to increase in number, while weaker species become scarce. Stronger species experience more intense intraspecific competition, because there are more of their own kind competing for common resources. Weaker species experience more intense interspecific competition, because they must compete with stronger, more successful species for resources. Weaker species, however, may evolve and improve their competitive abilities, whereupon they could increase and become abundant. These oscillations will continue until relative stability is reached between competing species.

III. Activity

In this chapter, we will examine competition between plants in a woodland community. If a woodland community is not available in your region, a desert scrub or similiar community will work fine. Competition among plants is more conspicuous than among animals. Plants show the results of competition through their growth rates, strata, and flowering times. There is a significant correlation between plant size and density of plants in a community. Generally, larger plants have more space between them because they monopolize resources and suppress seedling growth.

There are simpler methods to estimate competition between plants in a community than a vegetational community map which plots every single organism. The **distance-to-nearest-neighbor** technique is used to describe the relationship between two organisms and the rest of the community. Botanists use this technique often because it is easily used with stationary organisms such as trees. The distance-to-nearest-neighbor technique is limited in studying animals because of their mobility and because of difficulty in finding the nearest neighbor. It can be used on stationary animals (i.e., barnacles), mounds, nests, or burrows. By classifying both spatial arrangement and size, the researcher can determine competition.

Ecologists who apply the distance-to-nearest-neighbor technique use many methods to determine from what organisms they should measure to obtain a representative sample. The line-transect method is good because it is easy to construct and the organisms surveyed are chosen at random.

IV. Procedure

<u>Materials</u>

      100-meter chain
      10-meter tape measure
      4 ft. wooden stake
      surveyers ribbon
      cloth measuring tape
      field guide to local trees
      random number chart
      data sheets

Method

1. Find a wooded community located within a homogeneous abiotic environment. Variations in topography, moisture conditions, or substrate type cause plants the same distance apart to grow at different rates, therefore making it hard to detect competitive influences. It is best if there is one dominant tree species and several other tree species obviously competing for resources.

2. The instructor should identify the dominant species and two other common species that appear to be directly competing for the same resources. Only these three species will be used in measuring. Record the species identified on data sheet 4.1, section A.

3. Break the class into two groups and lay out two 100-meter straight lines with the survey chains (it is fine if the two lines intersect). Mark each end of the line with a wooden stake and survey ribbon.

4. There should be an equal number of measurements for each of the intraspecific and interspecific pair combinations. If three species are measured, then the following combinations of organisms should be examined:

Intraspecific	Interspecific
A-A	A-B
B-B	A-C
C-C	B-C

5. Students should break into pairs, decide which tree combination they will be responsible for, and measure their assigned combinations. This will assure that each of the tree species pairs are adequetely measured. At least five measurements should be taken for each pair combination (for a total of 10 for the two groups), or more if time permits.

6. To choose the locations of the tree pairs, use the random numbers chart in Appendix B. Enter the chart from any point and systematically choose numbers from the chart between 0 and 100. These numbers will correspond to the distance from the beginning of the survey chain (i.e., if 43 is chosen from the table, go to the 43rd meter on the survey chain).

7. The tree of the desired species nearest each random point should be utilized as the first member of the tree pair. The distance between the first selected tree and its nearest neighbor of the desired species should be measured (center of trunk to center of trunk). If an undesired species grows between the two desired species, thus impairing the competition between them, this pair should not be counted. Prior to sampling, the instructor should determine a maximum distance between trees in a pair (trees further apart than this distance do not have competitive influence upon each other). If no match to the pair is found within this distance, do not use this pair.

8. Measure the diameter of each tree at breast height (dbh) in each pair. Record the two species, diameter of each species, and distance between the two trees on data sheet 4.1, section B.

9. In the classroom, compile data and do calculations on data sheets 4.2, 4.3, and 4.4.

Data Sheet 4.1

Name _____   Date _____

Group Number _____

A) Identify the three major species in your community:

Species A _____
Species B _____
Species C _____

B) Complete the following table:

First Species	Diameter	Second Species	Diameter	Distance

Data Sheet 4.2

Name _____   Date _____
Group Number _____

C) Pairs working with **intraspecific** competition: Calculate averages for your tree pair combinations using data from data sheet 4.1, and enter averages in the appropriate row of your group's table in section D. Pairs working with **interspecific** competition, go to data sheet 4.3, section E.

D) Enter your data on the master table on the chalkboard. Complete the following tables using class data.

GROUP 1

Tree Combination	Avg. size 1st tree	Avg. size 2nd tree	Avg. distance
A - A			
B - B			
C - C			
Total averages			

GROUP 2

Tree Combination	Avg. size 1st tree	Avg. size 2nd tree	Avg. distance
A - A			
B - B			
C - C			
Total averages			

Data Sheet 4.3

Name _____   Date _____
Group Number _____

E) When examining interspecific competition, the average sizes of
    different species cannot be compared because the differences
    may be due to genetics, and not competition. Therefore, one
    species' size must be rescaled to assume the same value as the
    other species. First complete all data (except the rescaled size)
    in your group's table below, then complete section F, (on data
    sheet 4.4) in order to determine the rescaled size, and put these
    values in the table below.

GROUP 1

Tree combination	Average size 1st tree	Rescaled size, 1st tree	Average size 2nd tree	Rescaled size 2nd tree	Average distance
A - B					
A - C					
B - C					
Total averages					

GROUP 2

Tree combination	Average size 1st tree	Rescaled size, 1st tree	Average size 2nd tree	Rescaled size 2nd tree	Average distance
A - B					
A - C					
B - C					
Total averages					

Data Sheet 4.4

Name _____   Date _____
Group Number _____

F) To rescale the trees, the average diameter of each species is obtained so a ratio can be created to make the three averages the same value. First, calculate the average diameter of each species (for your group only) and complete columns two and three, below, based on size. (There are two values in section D and two values in section E for each species.) Compute the rescaled value by using the formula in column four. Go back to your group's table in section E and multiply the average diameters by the appropriate rescaled value to obtain the rescaled diameters. Swap data with the other group so that both tables in section E are complete.

YOUR GROUP

Size	Species	Average diameter	Formula	Rescaled value
smallest (s)			(s)/(s)	1
middle (m)			(s)/(m)	
largest (l)			(s)/(l)	

V. Questions

1. Based on the averages of the two transects (group 1 and 2), is this sampling technique representative of the community?

2. Which species shows the highest level of intraspecific competition? Which species shows the lowest? How do you know this? Are these species the same in both groups?

3. Which pairs of species show the highest level of interspecific competition? Which pair of species show the lowest? Explain. Are these species the same in both groups?

4. Which type of competition (intraspecific or interspecific) seems to be the most intense? Explain.

5. How does the size of an tree influence its proximity to its neighbors? How does it affect density of a community?

6. How do the basic features of competition influence such ecological processes of community structure and function, niche breadth and overlap, and evolution?

VI. Suggested Readings

Brower, J.; Zar, J. and von Ende, K. Field and Laboratory Methods for General Ecology. Third edition. Wm. C. Brown, Publishers; 1990.

Chiras, D. D. Environmental Science: Third Edition. Redwood City, CA: Benjamin/Cummings Publishing Company; 1991.

Cohen, J.C. Food Webs and Niche Space. Princeton, N.J.: Princeton University Press; 1978.

Cox, G.W. Laboratory Manual of General Ecology. Sixth edition. Wm. C. Brown, Publishers; 1990.

Kaskel, A.; Hummer, P. J. Jr.; Daniel, L. Biology: Laboratory Experiences. Columbus, OH: Charles E. Merrill Publishing Company; 1985.

Lemon, P. C. Field and Laboratory Guide for Ecology. Minneapolis, Minn: Burgess Publishing Company; 1962.

Pielou, E.C. Ecological Diversity. New York, N.Y.: John Wiley & Sons; 1975.

Smith, R.L. Ecology and Field Biology. New York, N.Y.: Harper & Row, Publishers; 1966.

Southwood, T. R. E. Ecological Methods: With Particular Reference to the Study of Insect Populations. New York, NY: Chapman and Hall; 1978.

Wratten, S.D.; Fry, G. Field and Laboratory Exercises in Ecology. Scotland: Edward Arnold (Publishers) Limited; 1980.

Chapter 5:
Community Succession

I. Objectives

This chapter will address the ecological importance of community succession. After completing the work, the student will be able to:

1. understand the processes and implications of community succession

2. observe successional changes through qualitative descriptions of vegetation in a field exercise

3. use quantitative analysis to compare successional stages involving hypothetical data

4. understand the importance of succession in creating and maintaining ecological communities.

II. Discussion

Communities are not merely static collections of organisms on a plot of land, but they are dynamic and constantly changing. Communities change in species structure, organic structure, and energy flow. Living organisms change abiotic factors, and in turn make their own niche intolerable. Newly created conditions allow for the invasion of competitor species. The changing of communities through time is called **succession**.

A **sere** is the sequence of communities in one area that develops during succession. Each individual community in the sequence is called a **seral stage. Primary succession** begins in areas that were previously devoid of life (i.e., rocks, cliffs, receding glaciers, and sand dunes). **Secondary succession** begins in areas that previously supported life, but were disturbed by human activity or natural conditions (i.e. abandoned farm land, construction sites, fire, or flood). The community is continuously modified until organisms that can tolerate the conditions they created can live in that community. These species can replace themselves without further modifying their habitat. The community becomes stabilized and is called the **climax community**. A true climax is rarely reached because there are always subtle changes in a community.

Organisms that inhabit different stages in community succession must meet different requirements. **Pioneer** species must be able to grow in areas low in nutrients and organic matter. They must be able to withstand wide variations in temperature, moisture and sunlight. As succession proceeds, habitat conditions change. Biomass, organic matter, and soil fertility all increase. There is a reduction in temperature extremes and fluctuations. The whole community becomes more complex. Increased habitat stratification, zonation, and niches, results in greater species diversity. As the community approaches its climax, the rate of change and the changes in species structure slows down. Succession changes a once-barren environment into a steady-state community filled with a variety of organisms.

Succession will continue on a relatively steady course unless it is disturbed in some way. Disturbed communities revert back to earlier, simpler stages. Natural occurrences such as fire help determine community structure. Humans, however, are the greatest cause of community disturbance. We influence ecosystem structure by cutting down forests, by allowing animals to graze, and by cultivation. In all cases, ecosystems are kept in the successional stage most desirable for our economic purposes; most of the time these are earlier stages. For example, cultivated areas are highly-artificial communities which require intensive care in the form of pesticides, herbicides, irrigation, and fertilizers. A monoculture crop is a far cry from the natural, diverse ecosystem that would normally be present. Our climax communities are cities. Industrialization and urbanization replaces natural vegetation with concrete, asphalt and steel. Only those species hardy enough to live in cities (cockroaches, rats, certain weeds) will thrive.

III. Activity

The study of succession is the study of habitats over time. Change is often slow and complete succession may take hundreds of years. Therefore, to study succession, ecologists must study several similar habitats in different stages of seral succession in order to get a view of the whole process. There are two ways to study community succession: qualitatively and quantitatively. A qualitative study involves going into the field and observing vegetation in various successional stages, paying attention to changes in stratification, coverage, and screening efficiency. Ecologists locate a disturbed community and one that has been left alone for a time and compare the seral stages.

A quantitative successional study involves measuring dominant species, coverage, density, frequency, biomass, or productivity in several habitats. These factors are related to abiotic factors in the habitat such as light, wind, and temperature to determine which seral stage the habitats are in.

In this activity, you will qualitatively observe succession in your community by locating and comparing similar habitats that have been recently disturbed and others that have remained relatively undisturbed. You will observe which species invade the area soon after disturbance and which species dominate in later seral stages. Secondly, since a great deal of data is needed to accurately study succession through quantitative means, in this activity you will use a set of hypothetical data to plot successional change in a community on a graph. From this data, you can determine which species is dominant in each stage and you can also see the cycling of species through the stages.

IV. Procedure

<u>Materials</u>

data sheets
calculator
field guide

<u>Method</u>

1. In the field, locate several similar habitats in different stages of succession. For example, you may want to compare an undisturbed habitat, abandoned farmland, and a recently disturbed construction site. Use data sheets 5.1 through 5.3 to record your observations, and compare the habitats on data sheet 5.4.

2. In the classroom, calculate the relative density of each species in the hypothetical data chart on data sheet 5.5.

3. Plot the relative density versus successional stage for each species on data sheet 5.6. Answer the questions in section V.

Data Sheet 5.1

Name _____   Date _____
Group Number _____

A) Answer the following questions about the first habitat observed.

    1) Describe the degree and time of the most recent disturbance.

    2) Was this disturbance caused by nature or by humans? Explain.

    3) What are the most dominant species in this habitat?

    4) Is there high or low diversity of species?

    5) Describe the abiotic factors (i.e., sunlight, soil condition, moisture, wind, erosion, etc.).

    6) Do you think this habitat is in an early or late seral stage of succession?

    7) Do you think this habitat is stable? Explain.

58

Data Sheet 5.2

Name _____   Date _____

Group Number _____

B) Answer the following questions about the second habitat observed.

1) Describe the degree and time of the most recent disturbance.

2) Was this disturbance caused by nature or by humans? Explain.

3) What are the most dominant species in this habitat?

4) Is there high or low diversity of species?

5) Describe the abiotic factors (i.e., sunlight, soil condition, moisture, wind, erosion, etc.).

6) Do you think this habitat is in an early or late seral stage of succession?

7) Do you think this habitat is stable? Explain.

Data Sheet 5.3

Name _____   Date _____
Group Number _____

C) Answer the following questions about the third habitat observed.

1) Describe the degree and time of the most recent disturbance.

2) Was this disturbance caused by nature or by humans? Explain.

3) What are the most dominant species in this habitat?

4) Is there high or low diversity of species?

5) Describe the abiotic factors (i.e., sunlight, soil condition, moisture, wind, erosion, etc.).

6) Do you think this habitat is in an early or late seral stage of succession?

7) Do you think this habitat is stable? Explain.

Data Sheet 5.4

Name _____   Date _____

Group Number _____

D) Use your observations from data sheets 5.1 - 5.3 to answer the following questions comparing the three habitats:

1) Rank the habitats in order from least to most disturbed. Why do you put them in this order?

2) What factors (both natural and human-made) caused the most disturbance? Are these factors the same in all 3 habitats? Why or why not?

3) Which dominant species occur in all habitats? Which dominant species are unique to one habitat? Explain.

4) Rank the habitats in order from lowest to highest species diversity. Explain why you put them in this order.

5) Which habitats would seem to be the most susceptible to minor fluctuations in abiotic factors? The least susceptible? Explain.

6) Based upon your answers to the above questions, which habitat is the least stable? The most stable? Explain.

Data Sheet 5.5

Name _____   Date _____
Group Number _____

E)   Use the following hypothetical data to complete the table.

Successional stage	Species A		Species B		Species C		Species D	
	D	RD	D	RD	D	RD	D	RD
1	0		0		59		23	
2	13		0		25		46	
3	17		8		20		35	
4	44		15		5		12	
5	9		30		0		0	
6	4		23		0		0	
Total density		n/a		n/a		n/a		n/a

F)   Calculate the relative density (RD) for each species in each successional stage.

1)   First calculate **total density** for each species by summing the species densities (D) in stages 1 - 6.

2)   Next calculate the **relative density** (RD) for each cell:
     RD = density/ total density

3)   Fill in the chart with the calculated results.

Data Sheet 5.6

Name _____   Date _____
Group Number _____

G) Graph the relative density (RD) versus the successional stage for each species.

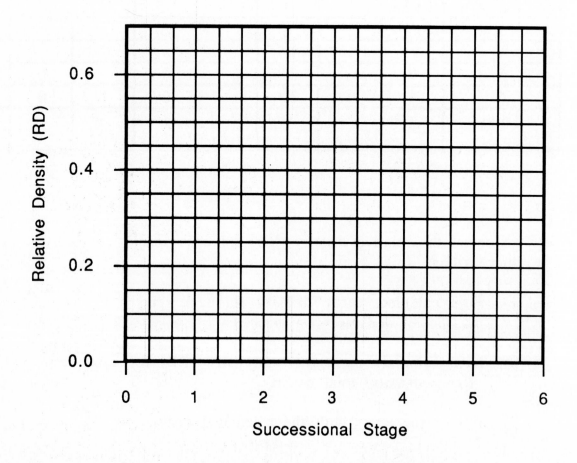

V. Questions

1. Rank the habitats observed on data sheets 5.1 through 5.4 according to successional order. Be sure to analyze all factors on data sheet 5.4 before making your decision.

2. Rank the species graphed on data sheet 5.6 in successional order.

3. Which species are the pioneer species? Which stage do you think is the climax stage and what species are present in this stage?

4. What do you think might have occurred between stages four and six? Justify your conclusion.

5. What role has human intervention played in ecological succession in your area? What are the implications for the environment?

VI. Suggested Readings

Brower, J.; Zar, J. and von Ende, K. Field and Laboratory Methods for General Ecology. Third edition. Wm. C. Brown, Publishers;

Chiras, D. D. Environmental Science: Third Edition. Redwood City, CA: Benjamin/Cummings Publishing Company; 1991.

Del Giorno, B.J.; Tissair, M.E. Environmental Science Activities: Handbook for Teachers. West Nyack, N.Y.: Parker Publishing Company, Inc.; 1975.

Gutierrez, L.T. and Fey, W. R. Ecosystem Succession: A General Hypothesis and a Test Model of a Grassland. Cambridge, Mass.: The MIT Press; 1980.

Nickelsburg, J. Field Trips: Ecology for Youth Leaders. Minneapolis, Minn.: Burgess Publishing Co.; 1966.

Smith, R.L. Ecology and Field Biology. New York, N.Y.: Harper & Row, Publishers; 1966.

Unit II

The Study of Populations

Every living organism cannot avoid being part of a population. Even though organisms act individually, they are also part of a larger whole (their species), which can be measured to predict the fate of that species. The survival of a species depends on how well it can (as a group) reproduce, utilize available resources, and adapt to a changing environment.

Populations are studied both from the aspect of ecological communities and from human demography. The same rules about populations apply to both the biological sciences and the humanities. Chapter Six addresses the underlying dynamics of population theory which are universal. Chapter Seven deals specifically with the implications of human demographic behavior.

Chapter 6: Population Dynamics

I. Objectives

Population dynamics addresses change in population size, composition, and distribution. After completing this chapter, the student will be able to:

1. understand the principle concepts of population dynamics

2. examine population growth using fruit flies

3. gain an awareness of the extent to which population is a relevant concept.

II. Discussion

Demography is the systematic study of a population. Three major demographic variables that affect the stability of a population are fertility, mortality, and migration. **Fertility** is the ability to reproduce. The general fertility rate (GFR) measures the number of births per year per 1000 females of reproductive age. The crude birth rate (CBR) measures the number of live births per year per 1000 population. Factors affecting birth rates include the availability of natural resources, space, competition, breeding habits, and male to female ratio. Human fertility rates are affected by a person's social, cultural, and economic environment.

Mortality is the number of deaths in a given time or place. High mortality is caused by famines and food shortages, epidemic diseases, competition, and depletion of resources. The crude mortality rate (CMR) is the number of deaths per year per 1000 population. Mortality problems specific to humans include lifestyle, diseases (cancer, obesity, and cardiovascular diseases), war losses, and poor living conditions due to poverty.

The **population growth rate** is a key measurement for demographers. The growth rate (in percentage) is calculated by subtracting the crude death rate from the crude birth rate, and multiplying by 100. An easy way to compare population growth is to calculate the **doubling time.** Doubling time is the time it takes for a species population to double in size. Doubling time is calculated by dividing 70 by the growth rate (in percentage). For example, the doubling time for some populations of microorganisms may be less than a day, while for humans the current doubling time is approximately 40 years.

Migration is the movement of organisms from one area to another. **Immigration** is movement into an area, and **emigration** is movement out of an area. Migration patterns can greatly affect population size and distribution. Organisms move for many different reasons, all of which increase their chance for survival. Animals migrate because of competition in their own species and between other species. Also, herd animals may follow their food resources with the changing seasons. Human destruction of habitat is a leading cause of animal migration. Humans migrate because they are in search of better economic opportunities or better climates. Oppressive governments force many people from their homelands involuntarily.

Two hundred years ago, Thomas Malthus noted that population growth increases exponentially, while food supplies increase arithmatically. This means that population growth will quickly outstrip food supplies unless population growth is held in check. If human population exceeds a region's **carrying capacity,** then they will have to migrate or they will die of starvation, disease, or war. People debate the actual carrying capacity of the earth because the carrying capacity is constantly being raised by advances in technology. People die of starvation and disease long before the carrying capacity is reached due to economic and social problems, not biological problems.

III. Activity

Fruit flies have long been used in scientific studies. They reproduce rapidly, have short generation times, and are cheap and easy to care for. This activity uses fruit flies to observe the consequences of rapid, unchecked population growth. You will observe colonies of fruit flies that had different initial numbers of flies, and count the number of offspring produced at the end of two weeks.

Once the fruit flies have been sedated with ether, they can be easily be divided by sex, and counted. Fruit flies can be sexed using the following information: female fruit flies are larger; they have pointed abdomens, with stripes extending to the tip of their abdomen. Male fruit flies are smaller, with rounded abdomens. The last third of the abdomen is solid black, not striped. Your instructor will be able to help you distinguish between the sexes.

For each colony, several calculations will be made in order to compare the numbers and growth rates of flies. A basic formula used in calculating exponential growth of a species is:

$$p_f = p_i * e^{rt}$$

where:
p_f = final population
p_i = initial population
e = a physical constant whose value is 2.7183
r = rate of growth
t = time

If you let $x = rt$, then the formula becomes:
$$p_f = p_i * e^x$$

Solving for e^x, we get:
$$e^x = p_f/p_i$$

Once e^x is determined, the corresponding value of x can be located, using the graph included in this chapter.

Knowing that $x = rt$, we can solve for r (in percentage), the rate of growth of that population:
$$r = x/t * 100$$

Instead of performing difficult calculations to obtain information on a species' population growth, you can easily determine a population's doubling time. Doubling time indicates how a species population is growing by quickly obtaining the time it takes for any population of that species to double, as the next equation illustrates. Once you know the rate of growth and time, you can easily estimate the final population number.

> We can determine the time it will take for that population to double in size by using the formula:
> $dt = 70 / r$, where r is a percentage

Once all information is calculated, you will be able to compare the growth rates of the colonies, and draw conclusions about population growth and its effects upon the environment and other organisms.

IV. Procedure

Materials

labeled jars of live fruit flies
ether
cotton balls
dissecting microscopes or magnifying lenses
white sheets of paper
paint brushes
calculators
data sheets

Method

1. Divide into groups of 2-4 (depending on the number of students in class).

2. Each group will obtain a colony of fruit flies that was prepared two weeks earlier.

3. Put a few drops of ether onto a cotton ball. (NOTE: BE EXTREMELY CAREFUL WHEN USING ETHER; DO NOT INHALE OR LIGHT A FLAME WHILE HANDLING!) Put the ether-soaked cotton ball over the top of the jar containing fruit flies **only** until the flies stop moving. Remove the cotton ball, and dump the flies onto the white sheet of paper (white paper makes flies easier to see).

4. Note the initial number of flies that had been placed in the bottle. Record this number of data sheet 6.1.

5. Carefully, but quickly (you don't want to take too long, or else the flies will wake up) divide the flies into female and male groups. Using the dissecting scope (or magnifying glass), count the flies of each sex, and record these numbers on data sheet 6.1.

6. Look in the bottom of the fly jars. Count all the dead flies you can. Record this number on data sheet 6.1. Dead flies will have their wings spread at a 90 degree angle away from their bodies; sleeping flies will have their wings folded close to their bodies.

7. Put your observations on the board for the rest of the class to record. Record the other group's information on data sheet 6.1, section B. Do the calculations on data sheet 6.2, and answer the questions in S-ection V.

8. Return the flies to their jars, and return the jars to your instructor.

Data Sheet 6.1

Name _____   Date _____

Group Number _____

A) Record the following:

1) Initial number of flies: _____

2) Number of males: _____

3) Number of females: _____

4) Total number of flies: _____

5) Number of dead flies: _____

B) Using information from the board, complete the table:

initial # of flies (p_i)	# males	# females	total # of flies (p_f)
2			
4			
8			
16			
32			

Data Sheet 6.2

Name _____   Date _____

Group Number _____

C) Complete the following table:

initial population pi	final population pf	e^x	x	time (t)	population growth rate (%) (r)	doubling time
2				14		
4				14		
8				14		
16				14		
32				14		
				14		
				14		

1) Determine the doubling time of each initial population:

a) divide the final population by the initial population and record this number in the e^x column
$$e^x = pf \: / \: pi$$

b) locate this e^x on the graphs on data sheets 6.3 and 6.4, and locate the corresponding x value; if e^x is between 0 and 15, use graph 6.1, if it is between 15 and 60, use graph 6.2. Record this value in the x column.

c) notice that **t** = 14 days

d) remembering that x = rt, solve for **r**, and record in the appropriate column.
$$r \: (\%) = (x \: / \: t) * 100$$

e) find the doubling time (**dt**) by dividing 70 by r (%), and record.
$$dt = 70 \: / \: r \: (\%)$$ (NOTE: use the value from step d; **do not** change it into a decimal)

Data Sheet 6.3

Name _____   Date _____

Group Number _____

D) Use the following graphs (6.1 and 6.2) to locate "x", as described in step (b) on data sheet 6.2:

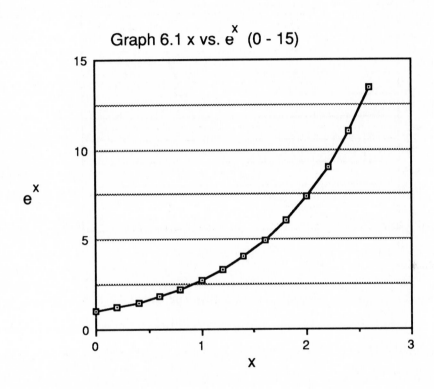

Graph 6.1 x vs. e^x (0 - 15)

76

Data Sheet 6.4

Name _____   Date _____

Group Number _____

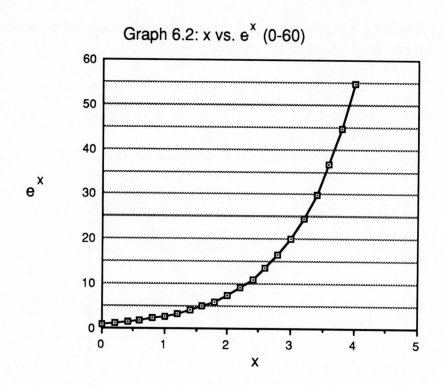

Graph 6.2: x vs. e^x (0-60)

V. Questions

1. For each of the colonies, were the number of females and males equal? Why or why not? What factors could account for unequal numbers?

2. For each of the colonies, was the population growth rate equal? Explain any differences in the growth rate.

3. In your opinion, which, if any, of the colonies approached that bottle's carrying capacity? Support your reasoning.

4. How can you apply what you have learned about growth rates, doubling times and their effects on the environment to human population growth?

VI. Suggested Readings

Chiras, D. D. Environmental Science: Third Edition. Redwood City, CA: Benjamin/Cummings Publishing Company; 1991.

Ginzburg, L. R.; Golenberg, E. M. Lectures in Theoretical Population Biology. Englewood Cliffs, NJ: Prentice-Hall, Inc.; 1985.

Menard, S.W.; Moen, E.W. . Perspectives on Population: An Introduction to Concepts and Issues. New York, NY: Oxford University Press; 1987.

Newman, J.L.; Matzke, G.E. Population: Patterns, Dynamics, and Prospects. Englewood Cliffs, NJ: Prentice-Hall; 1984.

Tuve, G.L. Energy, Environment, Populations, and Food. New York, NY: John Wiley & Sons; 1976.

Yaukey, D. Demography: The Study of Human Population. New York, NY: St. Martin's Press; 1985.

Chapter 7:
Human Population

I. Objectives

Human population dynamics addresses the mixed growth of global human population now and in the near future. Particular attention is paid to the way different nations endeavor to achieve zero population growth. Human population dynamics also studies the impact on resources used by humans over a period of time. After completing this chapter, the student will be able to:

1. understand the growth rate of different human populations by calculating doubling time and the associated current growth rates

2. examine the effects of projected decreases in growth rates over time with respect to reaching the goal of global zero population growth

3. study the impact of fertility on zero population growth

4. investigate the impact of population growth rate on consumption of global resources and sustainability of human societies.

II. Discussion

As presented in Chapter 6, the population growth rate of a species is the difference between the number of births and deaths in a given population. Generally, growth rate is expressed as a percentage. Knowing the growth rate, you can calculate a population's size for any time interval.

In table 7.1, the overall growth rate for the developed nations is 0.6 percent and for developing nations 2.1 percent. Therefore, the doubling time (dt = 70/r) for developed nations is 117 years and for developing nations is 33 years. The global population of our planet recently passed 5 billion. At the current population growth rates, in one year the populations of developed nations increased by 7.8 million people, from 1.3 to 1.3078 billion. During this same period, developing nations increased their population by 78 million people, from 3.7 - 3.778 billion.

Table 7.1 Growth Rate and Doubling Time in 1984		
Region	Growth Rate (%)	Doubling Time (years)
World	1.7	40
Developed Countries	0.6	117
Developing Countries	2.1	33
Africa	2.9	24
Asia	1.9	38
North America	0.7	99
Latin America	2.3	30
Europe	0.4	208
USSR	1.0	68
Oceania	1.3	55

Table 7.1 also shows that the doubling time for our planet is 41 years, which means that at the current growth rate, we will reach 10 billion around the year 2025. The consequences of this many people on our planet in less than 40 years from now would be devastating. The prospects of sustaining our species in a society as we know it today are dim. However, there are many choices we can make now that could provide a future that is not as bleak as projected if the historical population growth rate does not change.

Both crude birth and death rates affect the growth rate of a population. In order to bring the world to a condition of zero population growth, we need to understand and manage the factors that control crude birth rate. In addition, the factors controlling the crude death rate need to be understood. A lower crude death rate causes a population increase unless there is at least a similar decrease in the crude birth rate at the same time. To reach zero population growth, the crude birth rate has to decrease faster than the crude death rate until the growth rate approaches zero percent.

Using crude birth and death rates, we can gain an initial picture of a total population change by studying the manner in which populations grow. There are other tools we may use to further assess and predict the dynamics of population change. One of the most important is the **total fertility rate** (TFR) which is the number of children women are expected to have. When a nation has achieved zero population growth, the TFR reaches and maintains a value known as **replacement level fertility**. This is the fertility value at which couples produce enough children to replace themselves. In the developed countries, the replacement level fertility value is a TFR of 2.1 children. This means that ten couples must have a total of 21 children to replace themselves because, on an average, one out of 21 children die before reaching the age of reproduction. In the developing countries, the replacement level fertility is higher because the crude death rate is higher.

People mistakenly believe that when the TFR of a country reaches replacement fertility, the country has reached zero population growth. This is not true. Generally, women reproduce between the ages of 15 and 45. The population growth rate for a country will not start toward a rate approaching zero population growth until the number of women in their reproductive years decreases. For example, the TFR of the developed countries is 1.9 (1.8 for North America), which is less than replacement fertility (2.1), but the growth rate is still 0.6 percent (0.7 percent for North America). Even though each woman is having fewer babies, there are still many women having babies. North America is predicted to achieve zero population growth in approximately 70 years. If the TFR remains below 2.1 after the population growth has stabilized, then the population of that country may start to decrease.

The number of people on our planet intensifies the impact we have on the natural systems of the earth. The human impact on our planet, both regionally and globally, is determined by the available resources in a region and the stage of social development between subsistence based and industrial/high-technological societies.

In developing societies, food is a basic necessity constantly in high demand and low supply. Countries that are experiencing the highest population growth rates are also experiencing the worst food shortages. The highest population growth rates are in Africa and Latin America. If they cannot feed their present population, chances are they will not be able to feed future populations. Even though developing nations have high population growth rates, they do not have as high an impact per capita on their resources because their societies are not energy intensive.

Developed nations are characterized as having energy-intensive societies with low population growth rates. Developed societies critically impact a far greater range of global resources than developing countries. The long range problem facing our planet in achieving sustainability will be the degree to which we can manage the complicated relationships between populations and resources in the many different societies and cultures of the world. Presently, neither the developing nor developed countries are on paths that are sustainable. Each country has a unique set of problems in reaching sustainability; however, if we continue on the historical paths we are now pursuing, the result is nonsustainability for all. People are beginning to realize that it is not too late to reach sustainability - but we must begin now.

III. Activity

In Chapter 6, you studied population dynamics using fruit flies. Because they have a high reproduction rate it is easy to demonstrate their population growth rate and calculate their doubling time. In this chapter, you will be given historical data of selected populations and associated population growth rates. You will use the historical data and the basic formula from Chapter 6 to calculate the exponential growth of humans.

Remember, the relationship between an initial (p_i) and final p_f) population is:

$$p_f = p_i * e^{rt}$$

where: p_f = final population
p_i = initial population
e = a physical constant whose value is 2.7183
r = rate of growth
t = time

By defining x = rt, we can manipulate the formula to investigate several possibilities in human population growth. When you complete each section of the activity, you will be able to: compare human population dynamics in various parts of the world; compare how zero population growth is approached; and compare the factors that govern the impact a developed society makes on energy resources.

IV. Procedure

<u>Materials</u>

calculator
data sheets

<u>Methods</u>

1. Divide into groups of two.

2. In Sections A and B on data sheet 7.1, calculate and compare the population growth of developing and developed nations. Use table 7.1 in the discussion section for the required historical data.

3. Calculate the population of both the developed and developing nations if the growth rate decreases from 0.6 percent by 0.1 percent increments every 10 years for the developed world (Section C on data sheet 7.2); and from 2.0 percent by 0.4 percent increments every 10 years for the developing world (Section D on data sheet 7.2).

4. Calculate the energy consumption growth rate in the U.S. between 1965 and 1995 with the provided historical data through 1985 and projected data through 1995. Assume there is a constant population growth rate of 0.7 percent for the 3 decades of this period. Finally, we adjust total energy consumption for population growth by converting this data to energy (E) consumption per capita (Section E on data sheets 7.3 through 7.6). Use the information from the calculations to answer the questions in part V.

Data Sheet 7.1

Name _____   Date _____
Group Number _____

A) From table 7.1, fill in the appropriate data for growth rate and doubling time. Use $p_i = 3.7 \times 10^9$ (developing), and $p_i = 1.3 \times 10^9$ (developed) to determine the final population (p_f). Use the exponential equation below to determine p_f for each.

$$p_f = p_i * e^x, \quad \text{where } x = rt \text{ (change the rate from percent to decimal)}$$

B) Calculate p_f for the developing world using the developing world's rate of growth and the developed world's doubling time. (Use same equation as in section A).

	Region	r(%)	dt (years)	p_i (x 10^9)	p_f (x 10^9)
A	Developing World			3.7	
	Developed World			1.3	
B	Developing World		*	3.7	

*same time as developed world dt

Data Sheet 7.2

Name _____   Date _____

Group Number _____

Use the formula ($p_f = p_i * e^x$, where $x = rt$) to complete the following tables:

C) Calculate the final population (p_f) for developed nations where (r) starts at 0.6 percent and decreases by 0.1 percent every ten years until (r) = 0.0 percent. The final population becomes the initial population (p_i) for the next ten-year period.

Developed World			
r (%)	t (years)	p_i (x 10^9)	p_f (x 10^9)
0.6	10	1.3	
0.5	10		
0.4	10		
0.3	10		
0.2	10		
0.1	10		
0.0	10		

D) Calculate the final population for developing nations where (r) starts at 2.0 percent and decreases by 0.4 percent every ten years until (r) = 0.0 percent. The final population becomes the initial population for the next ten year period.

Developing World			
r (%)	t (years)	p_i (x 10^9)	p_f (x 10^9)
2.0	10	3.7	
1.6	10		
1.2	10		
0.8	10		
0.4	10		
0.2	10		
0.0	10		

Data Sheet 7.3

Name _____   Date _____
Group Number _____

E) Read the following explanation.

1) We will compare the impact of population on USA energy consumption from 1965 to 1995. The final comparison can be made in the form of a per capita energy use for each decade. The energy use is given in the table on data sheet 7.6. The U.S. population in 1985 was near 240 million. Using an average population growth of 0.7 percent over the three decade period, it is possible to estimate the population at each decade.

2) The annual energy consumption when expressed in BTU's per year is a large number. In 1965 it was 53×10^{15} BTU's. An energy unit based on 1×10^{15} BTU's is defined as one QUAD (from quadrillion). Therefore, the energy use for 1965 was 53 QUADS.

3) Using data from steps (1) and (2), we will compare the energy per capita use for each decade.

88

Data Sheet 7.4

Name _____   Date _____
Group Number _____

F) Complete the following exercises to determine the per capita energy consumption in the U.S. from 1965 to 1995.

    1) Find the population for 1965, 1975, and 1995 using the following formulas. Record your results on data sheet 7.6.

        a) p_f (1985) = p_i (1965) e^x

            $x = rt = (0.007)$ x   (20 yrs)

            $x =$ _____

            $e^x =$ _____

            p_i ('65) $= \dfrac{p_f\,('85)}{e^x} =$ _____ x 10^6

        b) Calculate p_i ('75).

            p_f ('85) = p_i ('75) e^x

            $x = rt =$ _____

            $e^x =$ _____

            p_i ('75) $= \dfrac{p_f\,('85)}{e^x} =$ _____

        c) Calculate p_f ('95).

            p_f ('95) = p_i ('65) e^x

            $x = rt =$ _____

            $e^x =$ _____

            p_f ('95) = _____

Data Sheet 7.5

Name _____   Date _____
Group Number _____

2) Calculate the energy consumption per capita (state units as BTU's/person) for each decade.

E = Energy
E_Q = Energy in QUADs (see table on data sheet 7.6)
E_c = Energy consumption

a) Calculate the energy consumption (E_c) per capita for 1965.

$$\text{per capita energy consumption} = \frac{E_Q \times 10^{15} \text{ BTU's ('65)}}{p_i \text{ ('65)}}$$

E_c per capita ('65) = _____ BTU's/person

b) Repeat step (a) for per capita E_c ('75) and E_c ('85). Record data in the table on data sheet 7.6.

c) If the energy consumption and growth rate of the first decade, 1965 - 1975, continues on to 1995, calculate the energy use in 1995 and the per capita use.

1) $E('75) = E('65) \, e^{rt}$

To solve for the growth rate, r:
$$\frac{E('75)}{E('65)} = e^x = \underline{\hspace{2cm}}$$

Using the graph on data sheet 7.7, x = _____
then $r = \frac{x}{t} = \underline{\hspace{2cm}}$

Data Sheet 7.6

Name _____   Date _____
Group Number _____

    2) Now we can find $E('95)$ between 1965 and 1995 using a constant rate.

        $E('95) = E('65)\ e^{rt}$
        $x = rt =$ _____
        $e^{x} =$ _____
        $E('95) = E('65)\ e^{x} =$ _____ QUADs

G) Based on your calculations, complete the table below.

Year	Energy (QUAD)	Population($\times 10^6$)	E_c (BTU)/person
1965	53		
1975	71		
1985	76	240	
1995			

Data Sheet 7.7

Name _____ Date _____

Group Number _____

H) Use the following graph to complete Section F, section 2), subsection c), part 1)

x vs. e^x

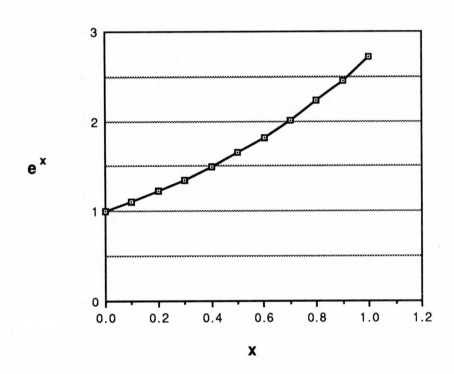

V. Questions

1. Considering the information derived in Section B on data sheet 7.1, do you believe the developing world's population will reach this number in 117 years? What percent of the world's total population would live in the developing world if the growth rates were to continue as they are presently for the next 117 years? Do you think the developing countries could adequately deal with this many people?

2. Comparing the data derived in Sections A, B, and D on data sheets 7.1 and 7.2, what would the population of the developing world be after approximately 70 years if there was no decrease in the population growth rate and if the growth rate decreases 0.4 percent every ten years? (Hint: double your answer for the developing world in Section A so that you have the population after 66 years.)

   Pop. of developing world = _____ with constant growth rate.
   Pop. of developing world = _____ with decreasing growth rate.

3. Which of these two population figures do you think is more realistic for the developing world in the year 2060? Do you think the lower of these two populations is sustainable on our planet?

4. What strategies may the developing countries start using right now to decrease the population growth rate?

5. Using the data you derived on data sheet 7.6, what is the trend in energy consumption per person in the U.S. between 1965 and 1995? What is rising at a faster rate, the consumption of energy or the population? Do you think the same trend would hold true in developing nations? Explain.

6. What do you think is most important for the developed world and the developing world respectively, to decrease the rate of population growth or to decrease the per capita energy consumption?

VI. Suggested Readings

Chiras, D. D. Environmental Science: Third Edition. Redwood City, CA: Benjamin/Cummings Publishing Company; 1991.

Ginzburg, L. R.; Golenberg, E. M. Lectures in Theoretical Population Biology. Englewood Cliffs, NJ: Prentice-Hall, Inc.; 1985.

Menard, S.W.; E.W. Moen. Perspectives on Population: An Introduction to Concepts and Issues. New York, NY: Oxford University Press; 1987.

Newman, J.L.; Matzke, G.E. Population: Patterns, Dynamics, and Prospects. Englewood Cliffs, NJ: Prentice-Hall; 1984.

Tuve, G.L. Energy, Environment, Populations, and Food. New York, NY: John Wiley & Sons; 1976.

Yaukey, D. Demography: The Study of Human Population. New York, NY: St. Martin's Press; 1985.

Unit III

Environmental Degradation

Environmental degradation of the ecosystem stems from the belief that global resources are unlimited. People believe that the earth is so large that the small amount of pollution they create individually will not come back to harm them. We perceive that the benefits of consumer products outweigh the risks of harm from pollution created during their production. In addition, cleaning up pollution created in the manufacture of a product costs more than people are willing to pay for that product. However, in achieving a sustainable society, the future costs of environmental degradation must be built into the present costs of production.

Chapter Eight describes the causes and implications of water pollution. Chapter Nine looks at who is creating air pollution and the extent of the problem. Both chapters discuss the implications of pollution upon human health.

Chapter 8:
Water Pollution

I. Objectives

After completing the work associated with this chapter, the student will be able to:

1. understand the implications of water pollution

2. perform a basic experiment to determine the water quality of a given system based on biochemical oxygen demand

3. gain an awareness of water problems in your own community and an appreciation of water conservation on a global scale.

II. Discussion

Water is the foundation of life on this planet. Without clean water we would soon dry up. Unless we quickly change our attitude about water pollution and water conservation, that may actually happen. Even though 70 percent of the earth's surface is covered with water, only three percent is fresh water, and less than one percent is usable.

Water pollution is defined as the release of substances into the environment that ultimately enter our water, in quantities that are harmful to organisms in the water or that use the water. It is hard to determine what level of water pollution irreversibly damages the environment. In order to save this resource for future generations, we need to err on the cautious side today.

The most obvious source of water pollution, called **point source pollution**, is directly discharged into water. Examples may include pipes coming directly from factories, open ditches, or sewage discharge. A less obvious source of pollution, called **nonpoint source pollution**, comes from a large, diffuse area, which includes air and land. Examples include pesticide and fertilizer runoff from farmland, urban runoff from storm drains, and acid deposition from air pollution. Nonpoint source pollution is more difficult to detect and treat than point source pollution.

Most water pollution can be divided between **organic** and **inorganic** pollutants. Organic pollutants are those that occur naturally, such as from sewage, agriculture, and food processing. Organic pollutants are usually not toxic themselves, but in large quantities they reduce the dissolved oxygen content of water. Inorganic pollutants are usually not biodegradable; they tend to stay in the system and become amplified through the food chain. Examples include acids, heavy metals, oil, phosphates, nitrates, and pesticides.

Major environmental effects of water pollution are reduction of dissolved oxygen, elevated levels of toxins, increased turbidity, and elevation of water temperatures. Potential ecological consequences include elimination of species, altered photosynthetic activity, and changes in community structure and function. We can reduce the impact of water pollution by improved treatment of waste before discharge, careful design and operation of systems, developing new technology, and tighter regulatory controls.

There have been many specific tests designed to test water quality. Most of these tests are specific for one type of pollutant and do not give a broad interpretation of water quality. In testing for organic pollution, a common method to determine general water quality is the Biochemical Oxygen Demand (BOD) test.

III. Activity

The BOD method indirectly analyzes organic content by measuring the amount of oxygen consumed by the bacteria living in the contaminated water over a period of time. Organic matter itself is not toxic in water. In low concentrations it may act as a fertilizer, but in larger concentrations it often produces toxic effects by encouraging the growth of bacteria and causing oxygen depletion. The higher the organic content, the higher the oxygen demand. This method measures how much oxygen is consumed through respiration and decomposition in a given sample.

Before we can measure the BOD of a water sample, we must know the dissolved oxygen (DO) content of the sample. The amount of DO in water affects natural biochemical processes in two ways: by limiting the amount of oxygen available for respiration, and by affecting the solubility of essential nutrients in the water. Four major factors that affect the dissolved oxygen content of water are (1) diffusion at the air-water interface, (2) the photosynthetic activity of plants, (3) the bacterial breakdown of waste products, and (4) the respiratory activities of other organisms in the water. In addition to these four, wind currents will affect the distribution of dissolved oxygen in open water.

The decomposition of organic matter is the greatest factor in the depletion of available oxygen in a given body of water. As the organic content of water increases, there is a corresponding increase in bacterial activity and a decrease in the dissolved oxygen content. This can be illustrated by a graph known as an Oxygen Sag Curve which plots, using arbitrary units, the dissolved oxygen concentration against distance or time from the point of discharge of the organic pollution.

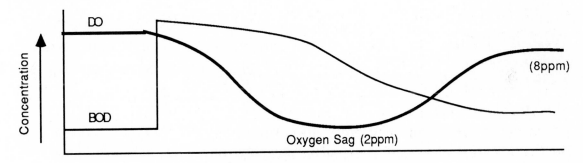

The dissolved oxygen content of the water is measured under saturated conditions and after incubating for five days in the dark at approximately 20° C. The difference between the two values is the amount used by bacterial decomposition and is known as the Biochemical Oxygen Demand (BOD).

$$BOD = DO_1 - DO_5$$

The dissolved oxygen content is measured using the azide modification of the BOD method. In this method a treated sample of water is titrated with a standard solution of phenylarsine oxide

(PAO) using starch as an indicator. Every milliliter of PAO used indicates the presence of 0.2 mg of dissolved oxygen. If only 100 ml of a sample is titrated, 1 ml of PAO would be equivalent to 2 mg of DO per liter. Therefore:

mg/l DO = ml PAO x 2

BOD provides a fairly reliable standard with which to compare the relative amounts of organic matter in an aquatic system. The acceptable degree of BOD loading in a particular aquatic system varies widely and can usually be determined by experience with similar systems. Remember, the cleaner an aquatic system is, the closer its BOD rating is to zero. The opposite is true for dissolved oxygen. An aquatic system is healthiest with a high amount of DO in the water. In this experiment, we will sample water from two different sites so we can compare them and determine if there is a difference and if either of them are dangerously polluted. Any sample with a BOD rating of 5.0 mg DO/L H_2O/5 days or greater should be further evaluated.

NOTE: This lab requires measuring DO twice: at day one, and at day five. If this lab will consist of just one meeting, the instructor will have to measure the first DO five days prior to the lab's regular meeting. The students will then perform the lab procedures from step 4, simply getting the DO data of day one from the instructor. Alternatively, the lab instructor may want the class to come in on their regular lab day and perform the day one DO measurements, and then come in on their own to perform the day five DO measurements. Finally, if time permits, the instructor may make this into a two-week course; the students would measure DO at day one, and seven, with little effect on accuracy.

IV. Procedure

C̲A̲U̲T̲I̲O̲N̲   H̲A̲Z̲A̲R̲D̲   C̲A̲U̲T̲I̲O̲N̲   H̲A̲Z̲A̲R̲D̲   C̲A̲U̲T̲I̲O̲N̲

THE CHEMICALS USED IN THIS EXPERIMENT CAN BE TOXIC, STAIN CLOTHES AND, IN THE CASE OF SULFURIC ACID, BURN HOLES IN CLOTHES AND SKIN. BE CAREFUL WHEN HANDLING REAGENTS. IF REAGENTS COME IN CONTACT WITH SKIN, FLUSH THE AREA THOROUGHLY WITH RUNNING WATER.

Materials

Each work station needs:

2 300 ml BOD bottles w/ ground glass stoppers
1 plastic cap for BOD bottle
3 200 ml Erlenmeyer flasks
1 10 ml buret
1 100 ml graduated cylinder
1 ring stand
1 buret holder

Each lab section needs one set of the following:

3 pipettes w/vacuum bulbs
2 one gal containers
   water sampler (optional)
2 three-foot sections of rubber tubing
   light-free incubator at 20° C
   squeeze bottles for PAO and starch solutions
   rubber gloves

Reagents

manganese sulfate ($MnSO_4$) solution
alkaline-iodide-azide reagent
concentrated sulfuric acid (H_2SO_4)
0.0250 N phenylarsine oxide (PAO) solution
starch indicator solution

102

Methods

1. Obtain water samples from two places you want to test and compare organic pollution levels (i.e., rivers, lakes, ponds, groundwater, storm drains, etc.).

2. Using the air outlet and rubber tubing, aerate the raw water samples for at least ten minutes to saturate with air.

3. Half of the class will test one water sample, and the other half will test the other. After obtaining raw data, the groups will exchange data. Two students should work at each work station.

    a) Each work station, using the rubber tubing, should siphon water into the bottom of their two BOD bottles, allowing them to overflow for at least ten seconds.

    b) Insert a stopper into the bottle and make sure there are no air bubbles trapped in the neck.

    c) Fill the neck of one bottle with water and cap with the plastic cap to prevent evaporation and air exchange. Place the bottle in the incubator at 20° C for five days.

4. Remove the stopper of the remaining bottle and add 2 ml of $MnSO_4$ solution to the sample using the appropriately labeled pipette. (<u>DO NOT</u> mix up the pipettes, as this will cause contamination of the reagents.)

5. Add 2 ml of alkaline-iodide-azide reagent in the same manner.

6. Put the stopper back in the bottle over the sink, pour off the excess, and mix by inverting 20 times. Allow the precipitate to settle until 1/3 of the liquid is clear. For the second time, invert 20 times and allow the precipitate to settle until 1/3 is again clear.

7. Remove the stopper and add 2 ml of concentrated H_2SO_4 by allowing the acid to flow down the neck of the bottle.

8. Put the stopper in the bottle, and rinse it under the faucet. Invert the sample again until the solution is thoroughly mixed. The treated water is now "fixed" and ready for titration.

9. Using a graduated cylinder, pour out 100 ml of the treated sample, then pour it into an Erlenmeyer flask. Repeat for the other two flasks so you have three 100 ml samples.

10. Titration: (for each Erlenmeyer flask)

   a) Fill the buret with PAO solution so that the starting level is zero.

   b) Titrate _slowly_ by allowing the PAO solution to drip into a flask (remember: do the complete titration process separately for each sample) until the water sample turns from _amber_ to _pale yellow_. The color change is more easily seen if the flask is set on white paper. Swirl the flask to assure that each drop of PAO is thoroughly mixed up.

   c) Add one squirt of starch solution to the flask. Starch merely turns the sample from pale yellow to blue, and serves as an indicator which allows for the visual determination of the titration endpoint. (Do not record the amount of PAO used yet!)

   d) Continue adding PAO (and swirling) slowly until the sample just turns clear.

   FOR BEST RESULTS RELEASE ONLY VERY SMALL AMOUNTS OF TITRANT (PAO) AT A TIME

11. Record the number of ml of PAO solution titrated on data sheet 8.1, table A.

12. Repeat steps 10 and 11 for the other two samples.

13. After five days, remove your other bottle from the incubator and repeat steps 4-12. Calculate the BOD in section B..

14. Obtain the raw data from the other sample location, fill in the table, and do calculations on data sheet 8.2.

Data Sheet 8.1

Name _____   Date _____

Group Number _____

LOCATION_____

A) RAW DATA:

DAY 1

Trial #	ml PAO used
1	
2	
3	
average ml PAO used	

DAY 5

Trial #	ml PAO used
1	
2	
3	
average ml PAO used	

B) DERIVED DATA:

DO_1 = average X 2 = _____mg DO/L H_2O

DO_5 = average X 2 = _____mg DO/L H_2O

BOD = DO_1 - DO_5 = _____mg DO/L H_2O/5 days

Data Sheet 8.2

Name _____   Date _____
Group Number _____

LOCATION_____

C) RAW DATA:

DAY 1

Trial #	ml PAO used
1	
2	
3	
average ml PAO used	

DAY 5

Trial #	ml PAO used
1	
2	
3	
average ml PAO used	

D) DERIVED DATA:

DO_1 = average X 2 = _____mg DO/L H_2O

DO_5 = average X 2 = _____mg DO/L H_2O

BOD = DO_1 - DO_5 = _____mg DO/L H_2O/5 days

V. Questions

1. Which of the factors that affect the oxygen content of water will add oxygen to the water, and which will reduce oxygen content?

2. Which of these factors were eliminated when you put your sample in a stoppered BOD bottle and placed it in a light-free incubator?

3. Which of the two test sites had the highest BOD rating? What would you conclude is polluting the water that made this test site have a higher rating?

4. Are either or both of your results dangerously high? If so, what can or should be done about the pollution?

5. Based on your experience with this lab, do you think water pollution is a problem in your area? In the nation? In the world?

VI. Suggested Readings

A Citizen's Guide to Clean Water. Arlington, VA: Izaak Walton League of America; 1973.

Addkison, R.; Sellick, D. Running Dry: How to Conserve Water Indoors and Out. New York, NY: Stein and Day, Publishers; 1983.

Chiras, D. D. Environmental Science: Third Edition. Redwood City, CA: Benjamin/Cummings Publishing Company; 1991.

Hellawell, J. M. Biological Indicators of Freshwater Pollution and Environmental Management. London, England: Elsevier Applied Science Publishers; 1986.

James, A.; Evison, L. eds. Biological Indicators of Water Quality. New York, NY: John Wiley & Sons; 1979.

Laws, E. Aquatic Pollution. New York, NY: John Wiley & Sons; 1981.

Lind, O.T. <u>Handbook of Common Methods of Limnology</u>. Saint Louis, MO: The C. V. Mosby Company; 1974.

Postel, S. <u>Conserving Water: The Untapped Alternative</u>. Worldwatch Paper, 67. Washington, DC: Worldwatch Institute; 1985.

Sittig, M. <u>Pollution Detection and Monitoring Handbook</u>. Park Ridge, NJ: Noyes Data Corporation; 1974.

Chapter 9:
Air Pollution

I. Objectives

After completing the work associated with this chapter, the student will be able to:

1. discuss how weather and topography influence air pollution and the effects air pollutants may have on local weather and global temperatures

2. list contributing factors and the major types of air pollution

3. discuss the significance of indoor and outdoor air pollutants

4. describe air sampling techniques for outdoor air pollution

5. determine whether their outdoor environment is free of air pollutants.

II. Discussion

Like water pollution, air pollution is becoming more of a problem in our society. It is obvious to anyone who has seen the skyline of a large city: Los Angeles, Denver, Houston, and Mexico City are already at the point of having to take action or subject their citizens to severe health risks.

Air pollution may be defined as the transfer of harmful amounts of natural and synthetic materials into the atmosphere as a direct or indirect result of human activity. The severity of air pollution in a given area is dependent upon climate, topography, population density, the number and type of industrial activities, and the organisms and materials affected.

The two major types of pollutants are **primary** and **secondary** pollutants. Primary pollutants are those occurring in harmful concentrations which are added directly to the air. They may be natural components of air present in greater than normal concentrations, or they may be something not usually found in the air. Secondary pollutants are those formed in the atmosphere through chemical reactions.

Most air pollutants are added to the troposphere, the lower layer of the atmosphere. In the troposphere, they often mix and react chemically with each other or with the natural components of the atmosphere. Eventually most of these pollutants and the by-products of chemical reactions are returned to the land or water by precipitation or fallout. Given the right physical conditions, insoluble and unreactive chemicals may diffuse upward into the ozone layer.

The major air pollutants are those produced in significant amounts and those having documented health and/or other environmental effects. They include: carbon oxides, sulfur oxides, nitrogen oxides, hydrocarbons, photochemical oxidants, and particulate matter.

Carbon oxides. Carbon monoxide (CO) is a colorless, odorless, tasteless, poisonous gas produced when any material or substance containing carbon, such as coal, oil, gasoline, or natural gas, is involved in incomplete combustion. Concentrations of CO encountered in heavy automobile traffic can cause headaches, a loss of visual acuity and decreased muscular coordination after several hours of exposure.

Even though carbon dioxide is a normal component of air, it is also an air pollutant. The amount present in the atmosphere is increasing due to the combustion of fossil fuels. As the amount of CO_2 increases, the temperature of the earth rises. Therefore, as global concentrations of CO_2 increase, the climate could be greatly affected.

Sulfur oxides. Sulfur dioxide (SO_2) and sulfur trioxide (SO_3) result from the combustion of solid and liquid fossil fuels containing sulfur. The sulfur joins with the oxygen in the air to form sulfur oxides. Emission sources include fuel combustion, metal processing, trash burning and chemical plants. Sulfur dioxide reacts with ozone, hydrogen peroxide, water vapor, and other substances to form sulfuric acid (H_2SO_4). Sulfur oxides can harm humans by irritating the upper respiratory tract and lung tissue. They can harm vegetation, corrode metal and stone, and cause a loss of tensile strength in natural and synthetic fibers.

Nitrogen oxides. Nitrogen oxides are emitted during fossil fuel combustion and some chemical manufacturing processes. Nitrogen dioxide (NO_2) is responsible for the yellow-brown color of smog, which greatly reduces visibility. Nitric oxide (NO) reacts with sunlight and hydrocarbons to produce photochemical smog. Exposure to high levels may cause lung irritation and increase respiratory disease. Nitrogen oxides also damage plants and may react with water to form acid rain.

Photochemical oxidants. Nitric oxides and hydrocarbons are primary pollutants which react in the presence of sunlight to produce photochemical oxidants (secondary pollutants). These secondary pollutants include nitrogen dioxide (NO_2), ozone (O_3), and peroxyacetyl nitrate (PAN, $CH_3CO_3NO_2$). Ozone is measured as an indicator of the presence of photochemical oxidants. Ozone is an odorous, colorless, poisonous gas. It can cause chest constriction, irritation of the mucous membrane, coughing, choking, headaches and severe fatigue. It can also damage many organic materials. Oxidants can also seriously damage plants, causing leaf lesions and reducing growth. At sea level, ozone is a pollutant, but at higher atmospheric levels it is a naturally occurring gas which protects the earth from harmful radiation. Chloroflurocarbons are pollutants which destroy the ozone layer and remove this protective screen.

Particulate matter. Particulate matter contributes approximately 5% of the weight of all air pollutants, and includes a variety of components such as sulfate salts, sulfuric acid droplets, metallic salts, dust, liquid sprays and mists, etc. Particles range in size from 0.005 microns to greater than 100 microns.

Natural sources of particulates are wind erosion, various pollens, volcanic eruptions and natural forest fires. These particles are usually large and settle out rapidly. Manufactured particulates usually come from one of three activities. **Attrition** adds particulates to the air by sanding, grinding, drilling and spraying. **Vaporization** is the phase change of substances from a liquid to a gas and is the major source of odors in the air. **Combustion** is the process of burning which results in the chemical combination of specific substances with oxygen. This is probably the major source of manufactured particulates. Incomplete combustion of fossil fuels produces heat, water vapor, CO and toxic or polluting substances. The major combustion sources are furnaces, internal combustion engines and incinerators.

Different sizes of particles have different effects on the environment and on human health. Very small aerosols tend to act as nuclei on which water vapor condenses. This increases the occurrence and durations of fogs and ground mists. Particle size also influences how deeply a particle may penetrate the human respiratory system. Larger particles are filtered out by nasal hairs then cleared by blowing the nose or washed to the gastrointestinal tract by mucous membrane action. Progressively smaller particles penetrate deeper into the respiratory system. Most are cleared from the lungs by the mucocilliary clearance mechanism. Particulates which reach the alveoli are often trapped and retained there indefinitely.

Indoor Air Pollution

In recent years many people have come to realize that air pollution problems are not limited to the outdoor environment. The average person may spend up to 90% of his or her time in indoor environments. Therefore, estimations of total personal exposure to air pollutants often correspond better to concentrations found in the indoor rather than the outdoor environment. As people have "tightened" and reduced ventilation in their homes and office buildings to increase the energy efficiency of the structures, they have created a situation where pollutants present in the building interior have no easy exit and therefore steadily accumulate to levels which often exceed those found outside.

Another situation which has contributed to the increased concern regarding indoor air quality is the growing number of building-related illnesses. "Sick-building syndrome" refers to a variety of health symptoms including irritation of the mucus membranes, headaches, dizziness, nausea, diarrhea, rashes, and abdominal and chest pains all associated with building occupancy.

Health effects of some pollutants can be extrapolated from existing health effects literature. At high concentrations these pollutants may have carcinogenic, allergenic, respiratory or other physiological effects. However, there is very little epidemiological data concerning the effect of most contaminants at the concentrations usually found in the indoor environment. Some pollutants may pose a health hazard only at concentrations above a threshold level, others may have no threshold. Pollutants proven to be carcinogenic are considered hazardous at any level. Synergistic effects may occur between pollutants or between pollutants and temperature and humidity. A synergistic effect occurs when the combined effect of two or more pollutants is either greater or less than the sum of their independent effects. There may be a latency period between exposure and manifestation of symptoms. This has contributed to the difficulty of obtaining data on health effects. Some latency periods may be more than twenty years. Often people who have been exposed to a pollutant will have moved from the area, may not realize they have been exposed, or may not connect present symptoms to an exposure occurring many years ago.

There are few air quality standards designed specifically for the indoor environment. Ambient air quality standards can be used as a guideline. In many areas, occupational standards are used. However, they are based on exposure conditions encountered during eight-hour shifts, six days a week. Exposure in the non-occupation indoor environment for some populations occurs 24 hours a day, seven days a week.

Control techniques for indoor air pollution include ventilation, source removal or substitution, source modification, air purification, and exposure avoidance by behavioral changes. The various control methods may be used separately or in combination. A list of the most common air pollutants is given in the Appendix.

Standards

The Environmental Protection Agency has established National Ambient Air Quality Standards (NAAQS) for air pollutants based on their effect on human health and the average exposure duration time. The standards are expressed in micrograms of a pollutant per cubic meter of air ($\mu g/m^3$). The pollutants should not exceed the standard of 150 $\mu g/m^3$ in a 24-hour period more than once a year, and are not to exceed an annual arithmatic mean of 50 g/m^3. A relative air pollution warning index has also been developed which describes the health effect caused by varying air quality levels.

The effects of air pollution may be ranked in six major classes:

Class One -- Nuisance and Aesthetic Insult
This involves odor, discoloration of buildings and monuments, and reduction in visibility.

Class Two -- Property Damage
Corrosion of metals is accelerated.

Class Three -- Damage to Plant and Animal Life
Air pollution contributes to leaf spotting and decay, decreased photosynthesis and crop yields. Animals, as well as humans, may suffer adverse effects to their respiratory and central nervous systems.

Class Four -- Damage to Human Health
Air pollution is most commonly associated with diseases of the bronchial tree. Air pollutants also irritate eyes and some pollutants buildup in the body and reach harmful levels. Some of the major diseases caused or aggravated by air pollutants are emphysema, bronchitis, bronchial asthma, and lung cancer.

Class Five -- Human Genetic and Reproductive Damage
Effects of air pollution on human genetics and the reproduction system are possible, though there is little documented evidence at this time.

Class Six -- Major Ecosystem Disruption
This class deals with the effects of air pollution on global and regional weather and climate and the resulting effects on ecosystems. It also deals with the direct ecosystem disruption caused by acid deposition.

There are several approaches to controlling air pollution. Input approaches involve changing the processes which produce air pollution, reducing fuel use and changing the type of fuel used. Output approaches involve collecting the pollutant before it is emitted into the atmosphere, shutting down plants during adverse meteorological conditions, and dispersing pollutants over a larger area to reduce concentrations.

III. Activity

In this exercise you will use a Hi-Volume air sampler to measure airborne particulate matter. This piece of equipment is similar to a large vacuum cleaner and is designed to pull air at a constant rate through a glass fiber filter. Particulate matter is trapped on the filter. The weight of particulate matter can be determined by weighing the filter before and after sampling. The filter should be handled only with forceps since dirt and oils from the hands would increase its weight. To eliminate any weight increase due to absorbed moisture the filter should be desiccated (dried) prior to weighing before sampling and after sampling. Since the Hi-Vol pulls air at a known rate, the volume of air sampled can be determined if the length of sampling time is known.

Most Hi-Volume samplers sample at a rate of 0.75 ± 0.25 cubic meter per minute (0.75 ± 0.25 m^3/min). Be sure that the Hi-Vol. sampler you are using is correctly calibrated. Using the raw data and a few simple equations you can determine the concentration of airborne particulates (C_p) present on the sampling days. This is done by calculating the weight of particulate matter per volume of air sampled. The C_p is expressed in μg/m^3.

In this lab, we will measure particulate matter over a four-day period. However, to complete all of the testing, it will require a time span of seven consecutive days. Each student is expected to come in one other day in addition to class time. The time schedule is explained further in the Procedure Section.

IV. Procedure

<u>Materials</u>

    forceps
4 glass microfibre filters (8" x 10")
4 envelopes
    desiccator
    Mettler balance
    Hi-Volume air sampler
    calculator
    data sheets

Methods

1. The instructions below explain each step in taking a particulate sample. Your instructor will demonstrate the procedure so that you can perform the appropriate steps on your day.

2. Number the filter appropriately (#1- #4) below the identification number with a pencil. Using forceps, carefully fold the filter into quarters and place it in an envelope. Now the envelope may be handled without contaminating the filter.

3. Place the envelope containing the filter into the desiccator for approximately 24 hours.

4. After 24 hours, remove the envelope from the desiccator and, using forceps, remove the filter from the envelope and weigh it on the Mettler balance. Record the initial weight (W_i) in the appropriate column on the table on data sheet 9.1.

5. Using forceps, remove the filter from the balance and place it, unfolded, numbered-side down in the frame of the Hi-Vol. Clamp it down, close the Hi-Vol and turn it on. Record the time on data sheet 9.1.

6. After 24 hours turn off the Hi-Vol, record the time, and remove the filter. Remember to use forceps. Place the filter back into the envelope, fold side down so no particulates are lost into the envelope. Desiccate for 24 hours.

7. After 24 hours, remove the envelope from the desiccator and, using forceps, remove the filter from the envelope and weigh it on the Mettler balance. Record the final weight (W_f) in the appropriate row and column on the table on data sheet 9.1.

8. Complete the calculations to determine the concentration of particulates (C_p). Record all data on data sheets 9.1.

9. Decide which day, other than lab day, you can come in; other students coming in that day will be in your group. Determine a time which you can all meet (remember, it should be approximately the same time as your lab period to ensure a 24-hour testing period).

10. Consult the table below to determine the activities your group needs to complete when it meets.

filter #	day						
	0	1	2	3	4	5	6
#1	dess.	weigh, hi-vol	dess.	weigh	-	-	-
#2	-	dess.	weigh, hi-vol	dess.	weigh	-	-
#3	-	-	dess.	weigh, hi-vol	dess.	weigh	-
#4	-	-	-	dess.	weigh, hi-vol	dess.	weigh

KEY: day 0 = day before class
day 1 = day of class
dess. = put filter in desiccator
weigh = weigh filter
hi-vol = mount filter on hi-volume sampler

11. The next week, in class, put your information on the board. Using other group's data, complete your chart on data sheet 9.1.

Data Sheet 9.1

Name _____   Date _____
Group Number _____

A) Complete the following table:

	filter #1	filter #2	filter #3	filter #4
initial weight (W_i)				
final weight (W_f)				
day and time the filter is put on the hi-vol sampler				
day and time the filter is taken off the hi-vol. sampler				
total time sampled (in min)				
mass of particulates ($M_p(\mu g)$)				
volume of air sampled (m^3/min)				
concentration of particulates (C_p)				

B) Complete the following calculations for each of the four filters:

1) To determine the mass of particulate matter collected (M_p), use the following formula:

$$M_p(\mu g) = (W_f - W_i) \times (10^6 \ \mu g/g)$$

2) To determine the volume of air sampled use the following formula:

$$Volume = time \ (min) \times machine \ flow \ rate \ (m^3/min)$$

3) To determine the concentration of particulates (C_p), use the following formula:

$$C_p = \frac{M_p \ (\mu g)}{volume \ (m^3/min)}$$

V. Questions

1. How is the severity of air pollution in a given environment dependent upon climate, topography, population density, and the number and type of industrial activities?

2. How might air pollution affect weather and climate? How might climate and weather affect air pollution?

3. Do you think indoor or outdoor air pollution is more of a health threat to people today? Explain.

4. Compare the results from your four samples. Describe any similarities and differences. What do you think may have caused variations from day to day?

5. Did your airborne particulate matter concentrations exceed the standard of 150 $\mu g/m^3$ in any 24-hour period? If so, what should be done about it?

6. Do you consider the air pollution in your area to be dangerous? If so, what are the contributors to air pollution in your area? If air pollution is not a problem now, do you think it will be in the future if nothing is done to protect your clean air?

VI. Suggested Readings

Airborne Particles. Subcommittee on Airborne Particles, Committee on Medical and Biologic Effects of Environmental Pollutants. Division of Medical Sciences Assembly of Life Sciences, National Research Council. Baltimore, MD: University Park Press; 1979.

Brunnée, J. Acid Rain and Ozone Layer Depletion: International Law and Regulation. Dobbs Ferry, NY: Transnational Publishers, Inc.; 1988.

Chiras, D. D. Environmental Science: Third Edition. Redwood City, CA: Benjamin/Cummings Publishing Company; 1991.

Elsom, D. Atmospheric Pollution: Causes, Effects and Control Policies. New York, NY: Basil Blackwell Inc.; 1987.

Gibson, M. To Breathe Freely: Risk, Consent, and Air. Totawa, NJ: Rowman & Allanheld Publishers; 1985.

Watson, A.; Bates, R.; Kennedy, D. eds. Air Pollution, the Automobile, and Public Health. Washington D.C.: National Academy Press; 1988.

Wellburn, A. Air Pollution and Acid Rain: The Biological Impact. New York, NY: Longman Scientific & Technical; 1988.

Unit IV

Conservation of Resources

We have become familiar with the function of ecosystems, and the impact of populations and pollution upon the environment. Now, we will study how to manage, conserve, and harness resources. Since humans are the biggest consumers of resources, we must be responsible for managing the earth's resources so that we can sustain ourselves and every other living organism.

Chapter Ten addresses the fact that without soil conservation we have no chance of feeding present and future populations. The issue of renewable energy sources is addressed in Chapter Eleven. We must find ways of perfecting and developing renewable forms of energy, and reducing our dependence for energy on nonrenewable fossil fuels. Finally, Chapter Twelve gives you an opportunity to learn more about current environmental problems by doing research.

Conservation of Resources

Chapter 10: Soil Management

I. Objectives

After completing the work associated with this chapter, the student should be able to:

1. measure the slope of an area

2. recognize, label, and describe the various horizons in a soil profile

3. classify a given soil sample based on its texture, density, pore space, and water-holding capacity

4. discuss critically the suitability of a given soil for agricultural cultivation or landfill location, etc.

II. DISCUSSION

Soil is defined as the unconsolidated mineral material located on the surface of the earth. Soil is formed by the physical weathering of parent material or bedrock. The parent material, through its chemical and physical properties, determines the type of soil that is eventually formed. The parent material can be weathered (split and crumbled) in several processes known as **physical weathering**, which include:

1. **freezing and thawing** -- this involves the expanding and contracting forces of water as it freezes and thaws.
2. **biological activity** -- the burrowing actions of animals and plant roots.
3. **friction** -- the grinding action of rock material due to gravity, wind, or water.

Another form of weathering is **chemical weathering**. Through many processes, such as dissolving of a solid, and reaction of substances with water, acids, or oxygen, chemical weathering causes the dissolution and weakening of rock. This results in fragmentation and leads to soil formation.

After soil is formed it may be transported from its site of formation by four methods. Soil is transported by water through the overflow of streams and rivers, the wave action of lakes, and direct flow over land during storms (runoff). Materials deposited by wind form deposits that often occur in times of drought. Gravitational forces cause downslope movement of soil. This can occur at a fast rate (mud slides) or a slow rate (soil creep). Soil material can also be transported by the movement of glacial ice.

The processes of soil formation and transportation are continually acting externally on the soil. At the same time, changes are occurring within the newly formed soil itself. These changes occur as the result of additions to the soil environment (through the use of fertilizers), losses from the soil environment (through leaching and erosion), transformation of material by chemical and biological reactions, and the translocation of materials to varying depths by water and soil organisms.

As a result of these processes, soil horizons develop. **Soil horizons** are the layers of soil in a given area which differ by physical structure as well as biological and chemical content. If one were to cut a section vertically downward through the soil, the various horizons would be visible. This vertical section is called a soil profile (refer to figure 10.1 on data sheet 10.2). Each well-developed soil has its own distinctive profile. These profiles are grouped by their dominant characteristics into soil orders. These characteristics are related to the nature of the horizons, the degree of weatherization, and the percentage of clay and organic matter.

The physical properties of soil include texture, structure, density, porosity, water, color, and biological impacts.

Soil Texture is determined by the proportion of soil particles of varying sizes, known as soil separates. The primary soil separates in decreasing order of coarseness are: sands, silts, and clays. The relative proportion of these separates determines a soil's texture. Once the percentage of clay, silt, and sand have been determined (based on the sizes of the soil separates), a textural triangle is used to determine the soil's textural class (Fig 2.). A soil's texture will affect water absorption and retention, aeration, and the fertility of the soil. Sandy soils drain well, are well aerated, and easy to till, but they also dry rapidly and lose vital nutrients as water drains through the soil. Clay soils have very small, tightly fitted particles. There is very little pore space and the soil is difficult to wet, drain, and till.

Soil structure is defined by the way different soil separates group together into a stable form or aggregate. Aggregates enhance the movement of air and water into and through the soil. Two types of soil aggregates are peds and clods. Peds are formed by natural cementing processes, while clods are formed artificially through plowing or digging. Three characteristics used to describe soil structural units are a soil's shape, size, and strength of cohesion.

Bulk density of a soil is the weight per unit volume of a soil as it exists naturally, including any air space and organic matter present. Bulk density is calculated from dried soil and does not contain water. A loose soil with large pore spaces will have less weight per unit volume than the same soil after compaction, which may result from the use of heavy machinery. Thus, bulk density can be used to estimate a soil's compactness or looseness. Bulk density can be used to determine the total water storage capacity of a given volume of soil and also to determine if the soil layers are too compact for root penetration and aeration.

Soil porosity is the property that enables air and water to circulate through soil. It is related to the total space in a given volume of soil not occupied by soil particles, but by air or water. Air space allows movement or spread of roots. A good root spread ensures the best possible supply of nutrients and water to a plant. A lack of water in pore spaces will inhibit plant growth and development, and will therefore increase the erosion of the soil. However, an excessive amount of water in pore spaces, known as oversaturation, will inhibit plant growth. This would also lead to overland storm runoff and soil erosion.

Soil water, or the amount of water in a soil, is important for plant growth and soil stability. Water-retaining forces are important properties of soil water. These forces are adhesive bonding, cohesive bonding, and hydrogen bonding. The result of these bonding forces is to hold water in soil. This bonding is a function of surface attraction; the larger the surface area per unit volume (i.e., smaller size particles), the greater the water-retaining force.

The strength with which soil holds water is important to plant growth. Water in clay soils is tightly bonded and is not the most suitable medium for plant growth. Sand is not a suitable medium because its large size has a smaller surface area to volume ratio and does not retain much of the water absorbed by the soil. An even mix of clay, silt and sand, known as loam, is the most suitable soil type for plant growth.

A difference in **soil color** between adjacent soils can indicate a difference in soil development or mineral origin. White colors indicate the presence of salts or carbonate deposits in the soil. Mottled colors indicate that a soil has periods of poor aeration annually. Blue, grey, or green tinged subsoils indicate long water-logged periods resulting in poor aeration. Dark colors usually denote soils containing high amounts of organic matter.

Soil organisms (macro and micro) aid in aeration, aggregation, organic decomposition, and nutrient distribution. Macro-organisms include plants, burrowing animals, insects and earthworms. Some micro-organisms such as bacteria, algae, fungi and protozoa fix nitrogen from the atmosphere (which is unavailable to plants) and release it in available forms to plant roots. Micro-organisms also increase the surface area and water-retention ability of the soil.

The soil's organic content increases its water-retention forces and nutrient levels. Organic matter also lowers the rate of soil erosion, encourages soil aggregation and provides for better aeration and water movement in the soil. Organic matter can occur naturally in a soil as detritus (decomposed plant and animal material), or it can be added to a soil in the form of composted material.

People usually degrade their land because they don't understand soils and how they act under various conditions. Through their ignorance, humans make poor planning decisions, which have long-lasting, negative effects on the land. In rural settings, farmers remove the total biomass, thereby degrading the soil, reducing its nutrients, and increasing erosion. Heavy farm equipment compacts soil, reducing its air and water-holding capacity. Finally, inorganic fertilizers and pesticides can build up in soils, and disturb their natural balance. In urban settings, new construction removes most vegetation, which results in increased erosion. Homes, factories, roads, and landfills reduce the availability of land for food production.

Erosion is a natural process which moves soil from one location to another by wind, water, or other natural action. The actions of humans accelerate this problem to great proportions. Erosion is a problem in countries like the United States, because there has been such an abundance of natural resources that people do not concern themselves with soil conservation practices. However, with ever-increasing pressures placed upon the land, people are realizing that our resources are not unlimited. In Third World countries, poverty and overpopulation drive people to use the land beyond its capacity for recovery in order to eat for one more day. If this degradation continues, what will be left for our children?

The goals of soil management are to prevent erosion and nutrient depletion. Farmers can prevent most erosion by using minimum tillage, contour farming, strip cropping, terracing, etc. Nutrient depletion can be prevented naturally by using crop rotation, organic farming, and polyculture practices.

Soil is a resilient resource and can withstand much abuse from human beings. Soil can degrade our pollution by physically trapping solid impurities, reacting chemically to absorb or dissolve impurities, and acting biologically to decompose organic wastes. While we use the earth as a receptacle for our waste products, we must not forget that the soil is of great value to us because we must live on it and produce all our food from it.

III. Activity

In this activity we will go out to the field and observe soil characteristics of that environment. We will measure the slope of the area, take soil probe samples, and sketch the soil profile. We will observe characteristics such as soil color, texture, depth, and water-holding capacity. We will test the pH, nitrogen, potassium, and potash levels of the soil.

Determining the slope is extremely important in deciding upon the best land use of an area. The greater the slope, the more problems a developer is likely to have. Of course, some problems related to building on sloped areas may be solved, but at additional costs to the builder which will be passed on to the buyer.

Soil color can help you determine related soil conditions. Use the following chart in making determinations about your soil samples.

Topsoil Condition	Dark (Dark grey, brown, black)	Mod. Dark (Dark brown to yellow-brown)	Light (Pale brown to yellow)
Amount of organic matter	Excellent	Good	Low
Erosion factor	Low	Medium	High
Available nitrogen	Excellent	Good	Low
Fertility	Excellent	Good	Low

The depth of soil (down to bedrock) has a measured effect on plant growth and water holding capacity. Deep soil (over 42 inches) has excellent water storage and plant growth capabilities. Moderately deep soil (20-42 inches) has good plant growth and water storage, and shallow soil (20 inches and under) has poor water storage and plant growth.

IV. Procedure

Materials

    4   meter sticks
    4   centimeter rulers
    4   levelling instruments
    4   soil probes
    4   Sudbury soil test kits
        soil sample collecting containers
        data sheets

Methods

1. Go to an area of land where you would like to determine its most valuable use. Try to find an area where the soil profile is already exposed so that you get a better prospective of soil strata. Answer the questions on data sheet 10.1.

2. Break into groups of 4 students.

3. First, each group will measure the average slope of the area.

    a) Select a site that represents the average slope of the land.

    b) Place one end of a 100-cm stick on the slope. Hold the stick outright until it is approximately level. Place a levelling instrument on the stick and raise or lower it until the bubble is centered.

    c) Measure the number of centimeters between the free end of the stick and the ground.

    d) The number of centimeters equals the slope of the land in percent. Record the percent of slope on data sheet 10.1.

4. With the soil probe, each group will gather a soil sample. Immediately measure the depth of each layer. On data sheet 10.2, draw and label the soil layers and depth. Use Figure 10.1 for help in identifying the layers.

5. Divide the sample by its different horizons and put each in a separate container. Bring these samples back to the lab for further analysis.

6. Complete the soil characteristics chart on data sheet 10.1. Use the following characteristics to identify:

Texture: Moisten the soil and squeeze it between your thumb and forefinger.

If it feels:	It is:
gritty	sand
smooth, slick, not sticky	silt
smooth, plastic, sticky	clay

Water-holding capacity: Fill the funnel with a soil sample, and pour water into the funnel. Watch the rate at which the water passes through the soil.

If it passes through:	It is:
easily	sand
slightly	silt
very slowly	clay

Color: See the color chart in the Activity section, and record whether the sample is dark, moderately dark, or light.

Amount of organic matter: Look at each layer. Approximate the percentage of organic matter. Record whether the sample is high, medium, or low in organic matter.

7. Each group will determine the chemical composition (nitrogen, phosphorous, potash, and pH) of the top two soil layers. Follow the directions that come with your soil testing kits. If time permits, do the other layers. Record data on the chemical composition chart on data sheet 10.1.

Data Sheet 10.1

Name _____   Date _____
Group Number _____

A) Complete the following description of your soil study area.

   1) What is the topography of your study area (hilly, flat, rocky, etc.)?

   2) How is this land and surrounding land currently being used?

   3) What type of biomass is the soil currently supporting?

   4) Do there seem to be any soil-erosion problems in the area? If so, what kind, and what appears to be the cause?

   5) What is the slope of your study area? _____ percent

B) Soil Characteristics Chart

Horizon	Texture	Water Capacity	Color	Organic Matter

C) Chemical Composition Chart

Layer	Nitrogen	Phosphorous	Potash	pH
1				
2				
3				

134

Data Sheet 10.2

Name _____   Date _____

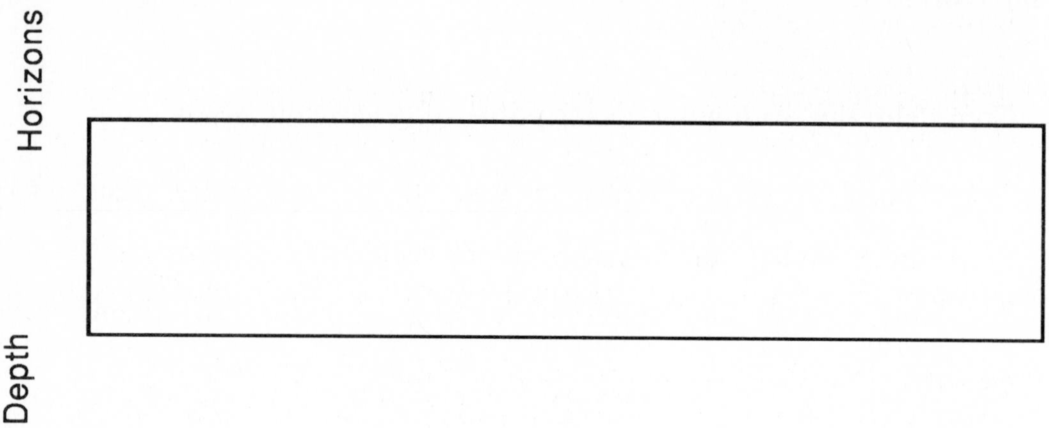

Horizons

Depth

Draw your own soil profile

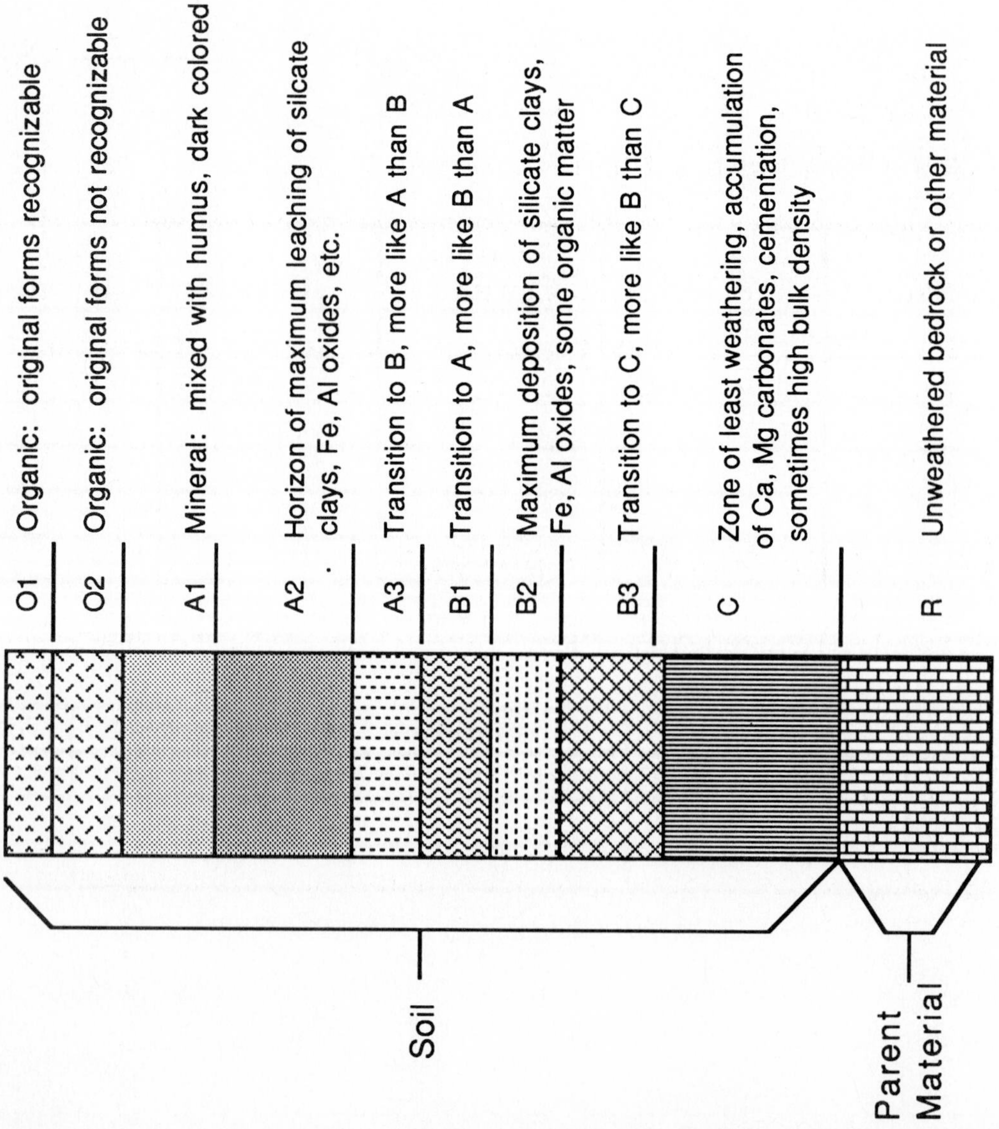

O1	Organic: original forms recognizable
O2	Organic: original forms not recognizable
A1	Mineral: mixed with humus, dark colored
A2	Horizon of maximum leaching of silicate clays, Fe, Al oxides, etc.
A3	Transition to B, more like A than B
B1	Transition to A, more like B than A
B2	Maximum deposition of silicate clays, Fe, Al oxides, some organic matter
B3	Transition to C, more like B than C
C	Zone of least weathering, accumulation of Ca, Mg carbonates, cementation, sometimes high bulk density
R	Unweathered bedrock or other material

Soil

Parent Material

Figure 10.1   A representative soil profile.

V. Questions

1. How will the slope of your soil survey area affect use of the land?

2. Is the soil texture in the area you surveyed best suited for cultivation or for a landfill site? (Hint: Is the texture mostly sand, silt, or clay?)

3. What other uses would the land be suitable for? Take into account soil type, environmental condition, and location.

4. Name one type of land use that is suitable for dark soil, moderately dark soil, and light soil (see the Activity section).

5. Does the nitrogen level in your top soil accurately depict what the soil color chart in the Activity section predicts?

6. From the chemical composition of your sample, what type of fertilizer would you recommend, if any, for your survey area if it is to be used for cultivation?

7. Based on the depth of your soil, is its water storage and plant growth capacity excellent, good, or poor?

8. What improvements to the survey area would you suggest to increase its use value and decrease the amount of environmental degradation?

VII. Suggested Readings

Biology/Science Materials. Burlington, NC: Carolina Biological Supply Company; 1988.

Chiras, D. D. Environmental Science: Third Edition. Redwood City, CA: Benjamin/Cummings Publishing Company; 1991.

Conserving Soil. Washington D.C.: United States Department of Agriculture, Soil Conservation Service; 1980.

136

Del Giorno, B. J.; Tissair, M. E. <u>Environmental Science Activities: Handbook for Teachers</u>. West Nyack, NY: Parker Publishing Company, Inc.; 1975.

<u>Forestry, Engineering and Environmental Equipment</u>. Jackson, MS: Forestry Suppliers, Inc.; 1990.

Jenny, H. <u>The Soil Resource: Origin and Behavior</u>. New York, NY: Springer-Verlag, Inc.; 1980.

Poincelot, R. P. <u>Toward a More Sustainable Agriculture</u>. Westport, CN: AVI Publishing Company; 1986.

Russell, J. E. <u>The World of the Soil</u>. London, England: William Brothers & Haram Ltd.; 1957.

Troeh, F. R.; Hobbs, J. A.; Donahue, R. L. <u>Soil and Water Conservation: For Productivity and Environmental Protection</u>. Englewood Cliffs, NJ: Prentice-Hall, Inc.; 1980.

Trudgill, S. T. <u>Soil and Vegetation Systems</u>. Oxford, England: Clarendon Press; 1977.

Chapter 11:
Renewable Energy

I. Objectives

This chapter will address the characteristics of renewable energies, the exchange of energy in the form of transfer of heat, and examples of extracting usable energy from the sun. After completing this chapter, the student will be able to:

1. understand the principles of transfer of heat energy

2. examine the storage of heat energy extracted from the sun

3. examine a method of transferring heat energy that has been stored

4. gain an awareness of the use of solar energy by humans.

II. Discussion

The countries of the developed world are very energy intensive. Indeed, these industrialized, highly technological societies could not exist without huge supplies of energy. The enormous growth in the consumption of energy by these societies is a major problem that we face. In Chapter 7, you saw that the increase in the use of energy in the U.S. was greater than the increase in population. Even though developing nations have a much higher rate of population growth, they use much less energy per capita than the developed nations. It will be difficult for the developing countries to evolve without obtaining large increases in the availability of energy.

Thus, both the developed and developing nations need energy. Until recently, the largest sources of energy have been from non-renewable resources (i.e., from fossil fuels), which are being rapidly depleted. Today, many people are involved in creating alternative methods of energy, in conserving and making more efficient use of energy, and in the large-scale development of renewable resources for all societies. Developed nations need diversify their energy base to produce and use various forms of energy in order to reduce their dependence on one energy source. Developing countries must expand production of their abundant, largely untapped energy resources. Both types of nations must begin to rely more upon renewable energy sources, and less on nonrenewables.

There are many renewable energy sources available for use on our planet. We can derive energy from the sun, wind, water, biomass, and through many forms of conservation. In developing energy resources, we can concentrate on a single source of available energy (i.e., solar energy in the desert), or we can utilize multiple sources of energy by integrating several systems (i.e., solar, wind, and biomass). The renewable energy resources that a country can develop depends on that country's location, the availability of various resources, and that country's economic condition.

In order to develop a strong energy base, we need to increase the availability and use of renewable resources. The most valuable and available renewable energy source we have is solar energy. Solar energy reaches the earth in various forms, one of which is heat. In order to use heat energy, however, we must transfer it from one form or place to another. There are three forms of heat transfer: conduction, convection, and radiation. The first two are involved in the heating and cooling of air and water, and represent a significant portion of the energy used in the developed nations.

Water is a good example of a substance capable of transferring heat energy from one source to another. Water can be heated by a number of energy sources, such as a fire, a hot water heater, or, as described in this chapter, the sun. Heat energy is transferred from the heat source (i.e., the sun) to the liquid (i.e., the water) by **conduction**, and is then moved through the water by **convection**. We can quantify this transfer of heat energy by defining it in terms of thermal units. For example, one type of thermal unit, the **calorie**, is defined as the amount of internal energy added to or extracted from one gram of water which will cause the internal energy to change one degree (+ or -) Celsius. We will use this simple relationship in the data section of the experiment.

III. Activity

In this lab, we will use the sun's energy to heat water. Common experience tells us there is heat energy in the sun: think of the thermal change in our skin as it is exposed to solar energy during the day, and the lack of such energy during the night.

We will be able to quantify changes in solar energy by measuring the change in water temperature caused by the energy from the sun. Water that is exposed to the sun will undergo a change in its internal energy. If we know the mass of the water and if we measure its change in temperature, we can determine the amount of the energy added to the water. The equation below demonstrates the relationship between the energy gained or lost (in calories), and the change in water temperature and the mass of the water:

$$E = (t_f - t_i) * m \text{ where}$$

t_i = initial water temperature (°C)
t_f = final water temperature (°C)
m = mass of the water in grams

In the first part of this lab, we will heat several containers of water using the sun as the energy source. We will measure the initial temperature of the water (before exposing it to the sun), and then measure the temperature of the water after exposing it to the sun's heat energy. The change in internal energy represents the amount of solar energy reaching the water and being absorbed by it. If the container is glass, then the solar energy that is absorbed is the portion of the solar energy that is transmitted through the glass. Changing the container in some way, (i.e. such as covering a part of the glass surface with aluminum, or using a black-painted metal can instead of glass) affects the solar energy transfer. When we use aluminum, we can either block the sun's rays, or concentrate them, depending upon how we position the container with respect to the sun. When we use a metal can, the metallic surface absorbs the heat energy, and transfers it by conduction through the metal to the water inside, where convection moves the heat through the water. The amount of energy added to the water is determined by the relationship of temperature and water mass. If the source of energy is removed (i.e. if the sun goes down), then energy will move from the walls of the container out to the surroundings until the temperature of the container and its surroundings is in equilibrium.

Energy from the now-hot water can be removed from the water by a heat exchange mechanism. We will use a copper pipe that is coiled and placed in the water. If water at a lower temperature is flowing through the pipe, the temperature of the water leaving the pipe will be higher than that entering the pipe, because the pipe will absorb heat from the hot water. Thus, the internal energy of the water in the pipe will have been increased. Likewise, the internal energy of the water in the container will have decreased. If there is energy still coming from the sun, the internal energy will not have decreased very much. If the extraction of the energy occurs after there is no further solar input, we can observe the relationship between the change in the internal energy of the stored system and the energy extracted.

In the later portions of this lab, we will measure the temperature differences between the tap water before it enters the coiled pipe, the water in the heating container, and the water exiting the pipe. We will manipulate the number of coils in the pipe, the speed of water going through the pipe, and the mass of water exiting the pipe, and determine if these factors influence the amount of heat energy transferred from the heated water into the water in the pipe.

In measuring the energy (in calories) of the heat transferred, if the calculations result in a negative energy, (i.e. if the final temperature is less than the initial temperature), this is fine, and simply indicates that energy left the system and was absorbed by some other system.

IV. Procedure

Materials

(Per Class)
3   one-gallon glass jars.
1   one-gallon can, painted black.
    aluminum foil

(Per Group)
1   of the above containers
    1000 ml graduated cylinder
1   lab thermometer.

1   coil of copper (brass) tubing (1/4 to 3/8 inch diameter), with coil comprising 6-8 turns.
1   coil of copper (or brass) tubing (1/4 to 3/8 inch diameter), with coil comprising 3-4 turns.
1   large (1000 ml) glass container
2   pieces of rubber tubing that fit the cold water tap and metal tubing.
    a place to expose the above containers to the sun at the same time.

Method

1. Divide the class into four groups. Each group will work with and take readings from one container.

2. Preparation of the containers:
    a) group **one** will use a clear-glass jar
    b) groups **two** and **three** will cover half of their glass jar with aluminum foil by taping the foil over half of the wall of the jar (do **not** put the foil completely around the circumference of the jar; cover half of the container's surface from top to bottom, leaving the other half clear glass)
    c) group **four** will use the black-painted can

3. Using the graduated cylinder, fill all the containers with **equal** amounts of water, and record the amounts on data sheet 11.1.

4. Measure and record the initial temperatures (t_i) of the water in each jar and record on data sheet 11.1.

5. Put the jars outside at the beginning of the day. Arrange the four containers in full sun as follows:

    a) clear glass jar: in the sun
    b) one aluminum-covered jar: put the foil-covered side opposite the sun so that the sun strikes the clear glass side.
    c) the other aluminum-covered jar: position the jar so that the sun strikes the foil-covered side first.
    d) black-painted can: in the sun

    (**NOTE**: If your lab is in the late morning, or in the afternoon, the lab instructor will do all the above steps, and will put the containers out in the early morning. When class time arrives, the class will divide into four groups, and each will be assigned one container)

6. Measure the final water temperature (t_f) of each jar at the start of class. Record on data sheet 11.1.

7. To measure the energy extracted from the sun, calculate the amount of energy added due to exposure of each container to the sun by using the formula on data sheet 11.1. Record your calculations on data sheet 11.1.

8. Place the 6-8 coil tubing in the heated water of your container ("heating container"). With your rubber tubing, connect one end of the coil to the cold water tap. Connect the second piece of rubber tubing to the other end of the coil, and put the free end of the tubing into the large glass container ("receiving container"). Turn on the tap, and **slowly** run water through the coil and into the receiving container until you get **400 ml** of water.

   a) Measure the temperature of the stored water in the heating container (t_i)
   b) measure the temperature of the water in the receiving container (t_f)
   c) Record the above on data sheet 11.2.

9. Determine the amount of energy transferred out of the storage containers and into the receiving container by the 6-8 coil tube, by using the energy formula on data sheet 11.2.

10. Repeat steps 8 and 9, but this time use the 3-4 coil pipe. Record and do calculations on data sheet 11.3.

11. In order to compare the amount of heat transferred when water is moving slowly through the coils to when it is moving quickly through the coils, proceed as follows:

   a) look back at data sheet 11.2, and copy the final temperature (t_f) that was measured in the receiving container for the 6-8 coil pipe when 400 ml of water was collected. Record this same number in Table 11.4 on data sheet 11.4.

b) remove the 3-4 coil pipe, and put the 6-8 coil pipe back into the storage container.

c) measure the temperature of the tap water (t_i), and record it in the 6-8 coil table for **both** 400 and 800 mls, on data sheet 11.4 (Tables 11.4 and 11.5).

d) turn on the tap so that the water is flowing through the coil at a faster rate than in step 8. Collect 800 ml of water in the receiving container.

e) measure the final temperature (t_f) of the 800 mls of water collected, and record it in Table 11.5 on data sheet 11.4.

12. Using the energy formula on data sheet 11.4, determine the influence of different rates of water speed, and mass, on the amount of energy transferred to the tap water from the storage container.

13. Repeat steps 11 and 12, this time using the 3-4 coil pipe. (For the 3-4 coil pipe, however, look back at data sheet 11.3, and copy the final temperature (t_f) that was measured in the receiving container for the 3-4 coil pipe when 400 ml of water was collected. Record this same number in Table 11.6 (t_f).) Record all other data on tables 11.6 and 11.7.

14. Write your results on the board, and get the results for the other containers. Record them on the appropriate data sheets.

Data Sheet 11.1

Name _____   Date _____
Group Number _____

A) Record the temperature (t_i) and (t_f) of the water in the containers.

Table 11.1: Initial vs. Final water temperature of storage container

	clear glass jar	aluminum side away from sun	aluminum side towards the sun	black-painted can
volume of water (ml)				
t_i				
t_f				

t_i = initial temperature of water
t_f = final temperature of water

B) Compute the solar energy added to the water:

1) to calculate the mass of water: 1 ml = 1 gm water
   m = _____ml water = _____gm of water

2) E = (t_f - t_i) * m
   =_____calories

146

Data Sheet 11.2

Name _____   Date _____
Group Number _____

C) Record the temperature (t_i) and (t_f) of the water in the storage and receiving containers.

Table 11.2: 6-8 coil pipe:

	clear glass jar	aluminum side away from sun	aluminum side towards the sun	black-painted can
volume of water (ml)				
t_i				
t_f				

t_i = temperature of water in the storage container
t_f = temperature of water in the receiving container

D) Compute the energy transferred from the heated water to the water inside the 6-8 coil pipe.

1) to calculate the mass of water: 1 ml = 1 gm water
   m = _____ml water = _____gm of water

2) E = (t_f -t_i) * m
   =_____calories

Data Sheet 11.3

Name _____   Date _____
Group Number _____

E) Record the temperature (t_i) and (t_f) of the water in the storage and receiving containers.

Table 11.3: 3-4 coil pipe:

	clear glass jar	aluminum side away from sun	aluminum side towards the sun	black-painted can
volume of water (ml)				
t_i				
t_f				

t_i = temperature of water in the storage container
t_f = temperature of water in the receiving container

F) Compute the energy transferred from the heated water to the water inside the 3-4 coil pipe.

1) to calculate the mass of water: 1 ml = 1 gm water
    m = _____ml water = _____gm of water

2) E = (t_f -t_i) * m
     =_____calories

Data Sheet 11.4

Name _____   Date _____
Group Number _____

G) Record the temperature (t_i) and (t_f) of the tap water and receiving containers.

Table 11.4: 6-8 coil pipe, slow-moving water:

	clear glass jar	aluminum side away from sun	aluminum side towards the sun	black-painted can
volume of water (ml)				
t_i				
t_f				

Table 11.5: 6-8 coil pipe, fast-moving water:

	clear glass jar	aluminum side away from sun	aluminum side towards the sun	black-painted can
volume of water (ml)				
t_i				
t_f				

t_i = temperature of tap water
t_f = temperature of water in the receiving container

H) Compute the energy transferred into the tap water from the storage container through the 6-8 coil pipe:

1) E (slow) = ($t_f - t_i$) * m
            = _____calories

2) E (fast) = ($t_f - t_i$) * m
            = _____calories

Data Sheet 11.5

Name _____ Date _____
Group Number _____

I) Record the temperature (t_i) and (t_f) of the tap water and receiving containers.

Table 11.6: 3-4 coil pipe, slow-moving water:

	clear glass jar	aluminum side away from sun	aluminum side towards the sun	black-painted can
volume of water (ml)				
t_i				
t_f				

Table 11.7: 3-4 coil pipe, fast-moving water:

	clear glass jar	aluminum side away from sun	aluminum side towards the sun	black-painted can
volume of water (ml)				
t_i				
t_f				

t_i = temperature of tap water
t_f = temperature of water in the receiving container

J) Compute the energy transferred into the tap water from the storage container through the 3-4 coil pipe:

1) E (slow) = $(t_f - t_i) * m$
= _____calories

2) E (fast) = $(t_f - t_i) * m$
= _____calories

V. Questions

1. In the first part of this activity, which container absorbed the most heat energy and reached the highest temperature? Which remained the lowest temperature? Explain the differing temperatures of the containers.

2. Since the mass of water was the same in each container, which appears to be the most efficient container in extracting solar energy? Why?

3. In transferring heat energy from the storage container, through the coils, and into the receiving container, was there any difference in the amount of energy transferred by the different containers?

4. Discuss the differences in the energy transferred by the 6-8 coil pipe and the 3-4 coil pipe. Which combination of container-type and coil number transferred the most energy? The least energy? Explain.

5. Discuss any differences in energy transfer between the slower-moving water and the faster-moving water. Did the doubling of water mass make up for the less amount of time the water was in the coils? Discuss any differences in coil size that may have affected the energy transfer between the storage container and the receiving container

6. Which combination of container, number of coils, water speed, and mass allowed for the greatest transfer of energy from the storage container to the receiving container? Explain.

7. Based upon what you have learned in this lab, describe the type of solar water heater (the type of collector, the number of coils, the water speed, and the mass of water) that would be most efficient in heating water, and/or transferring the heat energy to another water source.

VI. Suggested Readings

Chiras, D. D. <u>Environmental Science: Third Edition</u>. Redwood City, CA: Benjamin/Cummings Publishing Company; 1991.

Fisk, M. J.; Anderson, H. C. <u>Introduction to Solar Technology</u>. Reading, MA: Addison-Wesley Publishing Co.; 1982.

Stoner, C. H. <u>Producing Your Own Power: How to Make Nature's Energy Sources Work for You</u>. Emmaus, PA: Rodale Press, Inc; 1976.

Chapter 12:
Our Finite Resources:
The Current Affair

I. Objectives

Resource management problems are receiving more and more public attention. After completing this chapter, the student will be able to:

1. realize that most resource management problems are complex and poorly understood

2. understand that solutions to environmental problems will not be easily, or cheaply, reached

3. gain an awareness of what each person can do to minimize damage to the environment

4. gain experience in writing and presenting a research paper.

II. Discussion

Today you cannot watch the local newscast or pick up a newspaper without hearing or reading about environmental problems and issues on a local, national, and international level. We are quickly learning that our mineral and energy resources are not infinite, but will someday be depleted. In addition, the natural recycling of water and air is not sufficient to ensure the availability of these resources at the level of a population of 10 billion people. As we double our population every forty years we are putting increasing pressure upon the environment to meet our needs and desires. The infinite growth that entrepreneurs strive for cannot continue amidst finite resources.

Human beings must strike a balance between prosperity and sustainablity, if we are to remain a prominent species on the earth. The parameters of a sustainable ecosystem include recycling, efficiency, renewable resources, and zero population growth. Historically, humans have not worked within these limits. The challenge of this coming decade will be to reverse these trends and slow down the consumption of our natural resources.

Forecasters often disagree as to when our resources will be gone, but there is no question that it will happen. Should we deal with the problem now, or should we postpone action in the hope that new technology will provide a solution before our resources are gone?. Can we afford to leave our future existence up to chance? The economics of depletion reason that as a resource becomes scarce, it becomes more valuable. Therefore, the remaining amount must be conserved, or new, expensive technology must be developed to extract the remainder more efficiently. Our current society is slow to change. A more likely scenario may be that we will go on consuming resources at a ever-increasing rate until we reach a crisis. Through the hardship of our offspring, we may or may not find an alternative lifestyle. It is up to our generation to become motivated and ensure that this scenario does not happen.

III. Activity

Working in groups of 2-3 students, each group will research an environmental topic of their choice and present a 5-minute oral report on their research. Oral reports may be presented in debate style (with opposing views presented by the group), in reporting style, or any other style best suited to that topic. In addition, each group will turn in a typewritten paper, 3-5 pages long, covering its research.

Time will be given during this lab period to go to the library and start your research. However, extra time will have to be allowed to properly research and write this report. The report should be typewritten and double spaced. At least 5 sources should be used, with references properly documented at the end of the paper.

You may choose any topic from list A, or you may choose one of your own liking. Lists B, C and D include some of the best sources of recent information regarding many different topics concerning the environment. Talk with your librarian about specific material your school may have.

LIST A: Topics

Deforestation
    logging
    ranching
    peasant migration
    First World impact
    multi-national corporations
    old growth forest of North America
Energy
    alternative types
    renewable vs. nonrenewable forms
    projections for future uses
    conservation
    policy
Agriculture
    pesticides
    sustainable agriculture
    Third World dependency on First World economics
    the family farm
Conservation
    agriculture
    biological diversity
    habitat
        coral reef
        desert
        estuarine
        forest, etc.
    resources
        soil
        water
        air
Habitat Loss
    value: urban expansion vs. species extinction
Solid and toxic waste
    recycling
    Superfund sites
Environmental policy and laws
    political systems and resource management
    regulatory agencies and associations
    laws and Congressional acts
Career potential in environmental positions

Environmental ethics
   celebrities and rock musicians
   ecofeminism
Environment and religion
Environment and the military
Environment and technology
   environment vs. industry
   "Neo-Luddism"
Topics in the news
   global warming
   ozone depletion
   oil spills
   nuclear disasters, etc.

LIST B: Indices to Periodicals

Applied Science and Technology Index
Biological and Agricultural Index
Reader's Guide to Periodical Literature
Social Sciences Index
Government Repositories, located in many libraries

LIST C: Professional Periodicals

Advances in Environmental Science and Technology
Agricultural Water Management
Agriculture, Ecosystems, and Environment
American Journal of Alternative Agriculture
Aquaculture
Biological Conservation
Chemical & Engineering News
Environment and Behavior
Environment and Planning
Environmental Pollution
Environmental Research
Environmental Science & Technology
Foreign Policy
Journal of Energy Resources Technology
Journal of Environmental Economics and Management
Journal of Environmental Health
Journal of Environmental Quality
Journal of Environmental Sciences
Journal of Hazardous Materials

Journal of Social Issues
Journal of Soil and Water Conservation
Journal of the Institute of Energy
Monthly Review
New Scientist
New Statesman and Society
Organic Gardening
Political Studies
Power
Soil and Water Conservation News
Water Research

LIST D: General Periodicals

Audubon
BioScience
Current
Current Health
Discover
Environment
Field & Stream
Forbes
Futurist
International Wildlife
MacLean's
Mother Earth News
Mother Jones
National Geographic
National Wildlife

Outdoor Life
Psychology Today
Reader's Digest
Science
Science News
Sierra
Smithsonian
Society
Sunset
The Nation
Time
US News and World Report
USA Today
Utne Reader
World Health
World Press Review

Unit V

Environment and Society

In the first four units, we learned how biological communities function, how populations increase, the causes and effects of pollution, and the importance of conserving resources. But, learning something means nothing until you put it into action. The only way we are going to achieve global sustainability is if everyone becomes aware of our present course of destruction and alters their lifestyle.

This unit encourages participation in your local community. In Chapter Thirteen you learn to devise and administer an environmental survey. Chapter Fourteen helps you become aware of your community's infrastructure, how your community affects the environment, and ways you can effect changes in your community.

Chapter 13:
Environmental Survey

I. Objectives

After completing the work associated with this chapter the student will be able to:

1. construct, administer and analyze a survey

2. discuss the role surveys may play in environmental planning.

II. Discussion

Probably the most prevalent data collection technique in the social sciences is the survey. Surveys, and the data they generate, are important for environmental decision making. Land use planning, environmentally related political decisions, regulation of toxic waste disposal, creation of national parks and recreation areas, and the use of public funds for mass transit all depend upon information provided by surveys. The use and importance of surveys, however, is in no way restricted to environmental studies. Political surveys become an unending part of our lives each election year. Price surveys keep an eye on inflation. Market surveys determine possible demand for a new product or service. Journalists must be able to understand surveys in order to critically report them to the public.

Surveys are conducted in several ways. One may choose to interview with or without a prepared questionnaire, by mail, in person, or over the telephone. Or the objectives of the survey may dictate the researcher act as an observer, or search through records or card files. There are advantages and disadvantages to each type of survey. For example, mail surveys can decrease interviewer bias because, if a survey is conducted in person, the interviewer's facial expression or body language could affect the respondent's answer to a question.

Generally, however, mail surveys do not have a high response rate. This may be particularly important if response rate is related to some other variable, such as social or economic status. Although the initial mailings may represent an unbiased sample of the population to be questioned, the response obtained may not. For this lab, we will construct a written questionnaire to be administered by telephone.

Survey research has several strengths. It allows the researcher to gather large amounts of data on large numbers of people and is accurate within specifiable ranges of probability. Written questionnaires reduce interviewer bias and guarantee uniform question presentation.

Surveys also have several weaknesses. They may be very demanding of time and money; they are based on the assumption that respondents will give truthful answers; reliability and validity are difficult to check.

Surveys may be cross-sectional or longitudinal. Cross-sectional surveys conduct observation at one point in time. Longitudinal surveys conduct observations at two or more points in time. Cross-sectional surveys may be divided into two types: unweighted and weighted surveys. Unweighted cross-sectional surveys are those in which every element in the population has the same chance of being included in the sample. This is accomplished through random sampling. If some elements in the population are over-represented, the survey is weighted.

People who conduct surveys use their own set of vocabulary words to describe their methods and results. You must understand the following definitions in order to design and conduct a valid survey:

Sample - a subset of the population to be studied

Units of analysis - individuals to be included in the study; each member of the sample

Variable - a measurable characteristic of the unit of analysis

Independent variable - characteristic of the unit of analysis which is unalterable and may have an effect on the dependent variable

Dependent variable - variable which may change or vary under different conditions of the same independent variable

Population - all potential units of analysis defined by variables to be examined

Statistic - any summary measure of a variable of a sample

Parameter - summary measure of a population

Reliability - the extent to which the same procedure reproduced under the same conditions would yield the same results

Validity - the extent to which the procedure measures what it is intended to measure

Association - condition which exists between two variables when the distribution of one variable differs in some respect between at least some of the categories of the other variable; the way two variables relate to each other, discussed in terms of existence, strength, direction and nature.

EXAMPLE: In the study to determine how people in Waco feel about moving the Centex Zoo to Cameron Park, the population would consist of all Wacoans. The sample would consist of all those actually polled. Each individual of the survey is a unit of analysis. Variables might include sex, age, social status, income, number of times per year respondent visits the zoo, likelihood he or she will visit more often if the zoo is moved, willingness to accept a tax increase to finance the move, etc.

One statistic one might derive is that 43% of those polled would visit the zoo more often if it were moved. If every Wacoan were interviewed, the percentage represents a parameter. Independent variables might include sex, age, religious affiliation, or the number of children in the family. Dependent variables might include the likelihood a person would visit the zoo if it were moved. Marital status, the independent variable, might be found to influence the number of times a person visits the zoo every year, the dependent variable. If there is a difference between yearly visitations of those married and those unmarried, an association is said to exist between marital status and zoo visitation. Adequacy of a survey is discussed in terms of reliability and validity. If the survey were readministered under similar conditions, and the results were the same, the survey would be considered reliable. But if respondents were queried as to their willingness to finance the zoo through an increase in taxes when the final decision was to finance the move through an increased entry fee at the gate, the survey would be considered invalid.

III. Activity

Designing a questionnaire. Constructing a questionnaire is both an art and a science, usually involving five steps:

1. Specifying the informational objectives

2. Selecting the type of questionnaire

3. Writing the first draft

4. Pretesting the questionnaire

5. Revising the questionnaire

The word "questionnaire" is used to refer to either a personal interview schedule in which an interviewer asks questions and records answers, or to a self-administered questionnaire in which the respondent completes the task by himself. Questionnaires also may be distinguished by the structuring of the questions and responses. In close-ended questions, the respondent is asked to select his answer from a list of alternatives provided by the interviewer.

<u>EXAMPLE</u>: "What do you think is the most serious problem facing the U.S. today?"

    A) Racial relations
    B) Poverty
    C) Inflation
    D) The environment
    E) Unemployment
    F) Nuclear armament

G) Drugs

H) Other (specify) _____

The response categories should be both inclusive (include all alternatives), and mutually exclusive (an answer should be classified in only one category). Closed-ended questions are useful when the researcher knows enough about the topic to structure the alternatives.

Less structured items are called "open-ended questions." In open-ended questions, the respondent is asked to provide the answer to the question without reference to alternative answers.

Question wording. The phrasing of questions can be a very difficult process. Some of the more important guidelines for constructing questionnaire items are listed below:

1.   Questions should not be complex or double-barreled. The fallacy of the complex questions consists of asking for a "yes" or "no" answer to a question involving a number of parts. The counterpoint to this in survey research is called a double-barreled question.

Poor	Improved
Do you support environmental controls and clean air and clean water, or not?	Do you support environmental controls? Do you support clean air and water?

2.   Questions should not contain biased or emotionally loaded terms. Inclusion of terms with positive or negative connotations often influences the acceptance or rejection of a proposition.

Poor	Improved
Do you favor or oppose the child-killing bill?	Do you favor HB 211, which forbids voluntary termination of a pregnancy during the first trimester?

3. Questions should not seek technical information that the respondent cannot answer.

   Poor:
   Which do you think is more effective in curtailing the power of the oil companies: vertical or horizontal divestiture, price controls, or governmental regulations?

4. Questions should not be stated in the negative direction. Negative questions tend to confuse or mislead respondents who do not listen or read carefully.

   Poor   Improved
   Should marijuana not be   Should marijuana be
   decriminalized?   decriminalized?

Question sequencing. The order in which items are placed in the questionnaire influence the response to the question. If a closed-ended question precedes an open-ended one relating to the same topic, then the information contained in the closed-ended item is likely to influence the response to the open-ended item. Accordingly, open-ended questions normally precede closed-ended items.

Selecting the sample. The best way to accurately reflect the characteristics of a population is through random sampling. However, many times the population is too large to use this technique since a list of all members of the population is required. In these cases, the population may be divided into smaller groups from which random samples are taken. The random selection is carried out in such a way that every element in the population has an equal probability of being included in the sample.

Sample Survey-- people were asked the following questions on local parks

   Independent Variables

   Sex   (M/F)
   Number of children   (0, 1, 2, 3+)
   Marital status   (M/S)
   Age   (18-25, 26-40 , 41-55, 56+)

<u>Dependent Variables (questions)</u>

1) Approximately how often do you visit a local park?

Never_____   Seldom (1-3/yr)_____   Sometimes (4-8/yr)_____   Often (8-12/yr)_____   Frequently (13+/yr)_____

2) Which park do you most frequently visit?

Lake parks   _____
Central local parks   _____
Specialized parks   _____
Neighborhood parks   _____

3) When you visit a local park, what do you usually do?

Picnic   _____
Playground (children's)   _____
Organized sports (softball, golf, etc.)   _____
Pick-up sports (baseball, tag football, etc.)   _____
Exercise (running, aerobics, etc.)   _____
Simple relaxation (hanging out, etc.)   _____
Criminal mischief   _____
Felonious activity   _____

4) What would encourage you to visit local parks more often?

Improved facilities   _____
More security   _____
Change in climate   _____
More parkland   _____
Other (specify)   _____

5) I would support a small tax increase in order to fund park improvements.

Disagree   _____
Strongly disagree _____
Agree   _____
Strongly agree   _____

Recording your results. When initially recording your results, it is not possible to tally responses directly into a frequency table. Some intermediate recording technique is therefore required. One easy way is to record each individuals responses on a separate line of notebook paper. In the example below, three people have answered the questions from the above sample survey.

	Dependent Variables (questions)					Independent Variables			
person	ques. # 1	ques. # 2	ques. # 3	ques. # 4	ques. # 5	sex	# of child.	marital status	age
# 1	often	lake	picnic	more	agree	F	0	S	30
# 2	freq.	spec.	sports	more	agree	M	2	M	25
#3, etc.	sel	neigh	relax	secur	disag	F	3	M	50

When tallying the results of your survey, any method which seems logical to you may be used. However, you must record your data in such a way as to allow its compilation with the data gathered by the rest of the class. In other words, all of the respondents' answers (the units of analysis) must be tallied by both the question (dependent variable) and the independent variables.

The easiest way to accomplish this is to put your data into frequency tables. Please come to class with your data in tables. The following is a shortened version of the type of frequency table you should have. Make a table for **each** question. In the **top** half of each box, tally the number of responses for that box. In the **lower** half of each box, calculate what percentage that box is for the column. Note: your tables are for your use and do not have to be typed.

<u>EXAMPLE</u>: Question #1: <u>How often do you visit your local park?</u>

After completing the above table, you need to transfer your information from question #1 (in the dependent variable column), and all the independent variables to the frequency table for that question. In the above example, person 1 answered "often" to question #1. By combining the answer to question #1 with that respondant's characteristics (the independent variables), you can put tally marks in the boxes of each appropriate column (i.e., "often, female"; "often, 0 children"; "often, single"; **and "**often, 26-40 years old") So, for question #1, person 1, there are 4 tally marks put in the frequency table. Therefore, if 10 people answer your survey, you will have a total of 40 tally marks. When you complete your tallying, calculate what percentage each box is for the total of that column, and put that percentage in the lower portion of the appropriate box (i.e., if two females answered "seldom", three answered "sometimes", one answered "often", and four answered "frequently", the percentages **for that column** would be: 20% "seldom", 30% "sometimes", 10% "often", and 40% "frequently")

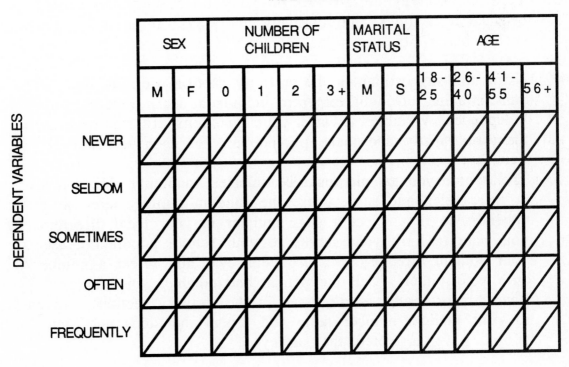

IV. Procedure

Materials

telephone book
random number table (see appendix)
calculator

Methods

1. The class as a whole will choose and design a survey. During the week, each student will survey ten people and make frequency tables for each question. The following week the class will choose two questions for which the class will tabulate the total results.

2. The class will choose a survey topic. Choose something that is an issue in your local community, preferably something to do with the environment. This can cover a wide range of topics such as public land use, pollution, city planning, environmental attitudes, ethics, or specific local issues. Be creative.

3. Your best survey design will probably be a telephone survey because it is quick and relatively unbiased. If another survey design better fits your topic, go with it and improvise.

4. Using the suggestions in the Activity section, design your questionnaire. Do not forget to include at least three independent variables. Once your questions are finalized, write them out neatly and make photocopies for each student.

5. The people you sample must be selected randomly. If you are doing a telephone survey, use the random number table in the appendix to first select a page number in your local directory. (If you want to limit your survey to your school population, use your student directory.) Close your eyes and select a number with your finger. Call the person and conduct your survey. Record your results as discussed in the Activity section. Go back to your random number table and repeat the steps until ten people have answered the survey.

6. Make a frequency chart for each of the questions, as described in the Activity section.

7. The next week in class, choose the two questions that appear to show the most correlation between the independent and dependent variables. On the chalk board, tabulate everyone's results and calculate percentages for these two questions. Make a copy of this for yourself on notebook paper.

8. When writing your report, analyze the percentages to determine the correlation between the independent and dependent variables. Include your questionnaire with your lab report.

V. Questions

1. Describe the results of your survey.

2. Which independent variable showed the greatest association to the dependent variables?

3. How would you improve the survey your class performed?

4. Were people responsive to your survey or did they show a lack of concern? Explain.

5. Were your results different than what you expected before you conducted the survey?

6. How could an environmental survey be used in community planning?

VI. Suggested Readings

Chiras, D. D. Environmental Science: Third Edition. Redwood City, CA: Benjamin/Cummings Publishing Company; 1991.

Jessen, R. J. Statistical Survey Techniques. New York, NY: John Wiley & Sons; 1978.

Kimble, G. A. How to Use (and Misuse) Statistics. Englewood, CA: Prentice-Hall, Inc.; 1978.

Loftus, G. R.; Loftus, E. F. Essence of Statistics. New York, NY: Alfred A. Knopf; 1988.

Mendenhall, W.; Ott, L.; Scheaffer, R. L. Elementary Survey Sampling. Belmont, CA: Wadsworth Publishing Company, Inc. 1971.

Slonim, M. J. Sampling in a Nutshell. New York, NY: Simon and Schuster; 1960.

Turner, C. F.; Martin, E. eds. Surveying Subjective Phenomena, vol. 1. New York, NY: Russell Sage Foundation, 1984.

Chapter 14: Environmental Awareness and Lifestyle

I. Objectives

After completing this chapter the students will be able to:

1. be aware of how their everyday actions affect the local environment

2. be familiar with changes they can make in their lifestyles to improve the protection and conservation of resources

3. discuss the water treatment, waste water treatment, power generation and solid waste disposal in their local community.

II. Discussion

Throughout this course we have discussed many of the components that make up the world we live in and how they interact with each other. You now know that we are a part of nature and not apart from nature. We are still a long way from achieving the sustainable society that we need in order to keep living on this planet. To achieve this, everyone will have to make changes in his or her personal lifestyle. Working as an individual is not enough; one must also work within the decision-making infrastructure of society in order to effect changes.

Sometimes environmental problems are so overwhelming and beyond control that one person's actions seem hopeless. But if many people come together, their actions will not be hopeless. There are many things that an individual can do to help the environment. The home is a good place to start. Utility use can be reduced in a variety of innovative ways. Water, electricity, and gas may all be conserved by retailoring personal use habits to use only what is necessary to get the job done. Entrepreneurs are constantly developing and marketing new technologies to increase the efficiency of household appliances.

The disposal of solid waste is an increasing problem around the world. There are several ways to decrease the amount of solid wastes produced in your home. Glass, aluminum, paper, and, increasingly, plastics can all be recycled. Organic wastes (i.e. yard clippings, and inedible foodstuffs) can be composted and turned into high-quality fertilizer. Gardening and wise shopping can greatly reduce the amount of packaged goods consumed in your home. In addition, fresh foods tend to be healthier. Multiple use of shopping bags and containers can increase the life span of products.

Americans consider owning an automobile practically a First Amendment right. By foregoing indiscriminate use of your car, you can reduce cost and pollution. Car pooling, public transportation, and reducing the number of trips will save energy and money. For trips under several miles, walking or riding a bicycle is healthier, cleaner, and more efficient.

On the community level, both awareness and action are important. Being aware of how your community functions, and problems associated with its functioning, are first steps in solving environmental problems. You must understand the dynamics of community utilities such as power production, water and sewage treatment, and solid waste disposal, before you can consider how to solve environmental problems associated with them.

To effect changes, you must understand the political infrastructure of your local community. The interplay of power between elected officials and the business community cannot be underestimated when addressing environmental issues. A major part of solving problems is dealing with the people involved with the issue and not just the ecological aspects of the issue. Until recently, short-term profit has superseded long-term environmental cost. Hopefully, people will realize that future costs must be built into present-day consumption.

Train yourself to be attuned to environmental problems by observing what goes on in your community. Pay attention to local and national media coverage. Do not passively accept things as they are, just because they have always been done that way. Learn to be objective when viewing the world around you.

Once you understand the problem, the next step is to become involved. Attend meetings, join groups, form committees, and write to elected officials and newspapers. Actively participating in your community will insure that your voice is heard and the environment will be preserved for future generations.

III. Activity

This activity will teach you to observe and question how your local environment is handled. Discuss how your local community works and how you can become involved in environmental issues. Identify community problems and possible solutions. Plan a field trip to one or more places, such as the following: water-treatment plant, sewage-treatment plant, solid-waste disposal unit, power generation plant, recycling facility, Federal or local government agency that controls environmental standards, or a town-hall or city-council meeting. Be sure to address each of the following questions in your laboratory write-up.

IV. Questions

1. What method is used for treating your drinking water? Your sewage? What happens to your treated sewage water?

2. How is your electrical power produced? Does your power plant produce pollutants? What kinds?

3. How are solid wastes disposed of in your community? Describe problems specific to your locality.

4. Describe specifically what you observed on your field trip.

5. Based upon your class discussion and observations from the field trip, tie together how your community is dealing with each of the previous unit topics in this course (ecology, population, resources, and pollution).

VI. Suggested Readings

Brehman T.R. <u>Environmental Demonstrations, Experiments and Projects for the Secondary School</u>. West Nyack, NY: Parker Publishing Company, Inc.; 1973.

Brower, J.; Zar, J. and von Ende, K. <u>Field and Laboratory Methods for General Ecology</u>. Third edition. Wm. C. Brown, Publishers; 1990.

Chiras, D. D. <u>Environmental Science: Third Edition</u>. Redwood City, CA: Benjamin/Cummings Publishing Company; 1991.

Del Giorno, B. J.; Tissair, M. E. <u>Environmental Science Activities: Handbook for Teachers</u>. West Nyack, NY: Parker Publishing Company, Inc.; 1975.

Kotsonis, H. H.; Baker, B. <u>Modern Lesson Plans in Environmental Science</u>. West Nyack, NY: Parker Publishing Company, Inc.; 1972.

Lemon, P. C. <u>Field and Laboratory Guide for Ecology</u>. Minneapolis, Minn: Burgess Publishing Company; 1962.

Wratten, S.D.; Fry, G. <u>Field and Laboratory Exercises in Ecology</u>. Scotland: Edward Arnold (Publishers) Limited; 1980.

Appendices

Appendix A: Commonly-Used Conversion Factors

Length

1 km	=	1,000	m
1 m	=	100	cm
1 m	=	1,000	mm

1 cm	=	0.01	m
1 mm	=	0.001	m

1 ft = 12 in
1 yd = 3 ft

1 mi = 5280 ft

1 km = 0.621 mi
1 ft = 0.305 m
1 yd = 0.914 m

1 m = 39.4 in
1 in = 2. 54 cm

Volume

1 ml = 0.001 L

1 ml = 1 cm^3

1 gal = 4 qts
1 qt = 2 pts

1 qt = 4 cups
8 oz = 1 cup

1 L = 0.265 gal
1 L = 1.06 qt
1 L = 0.0353 ft^3

1 barrel = 35.3 ft^3
1 barrel = 0.765 yd^3

Mass

1 kg = 1,000 g
1 mg = 0.001 g
1 ton = 2,000 lb

1 g = 1,000 mg
1 metric ton = 1,000 kg
1 lb = 16 oz

1 kg = 2.20 lb
1 lb = 454 g

1 g = 0.035 oz

Temperature

$$°C = \frac{(°F - 32.0)}{1.80}$$

$$°F = (°C \times 1.80) + 32.0$$

Appendix B: Random-Numbers Chart

62	2	36	27	36	30	30	63	78	46
35	78	67	68	39	41	52	5	31	52
74	20	96	82	28	29	6	99	42	27
88	2	78	8	59	34	2	18	59	34
63	41	90	6	36	65	1	98	64	99
97	51	22	56	78	40	1	25	69	38
60	4	86	33	28	78	41	78	78	5
69	27	44	48	67	66	78	98	55	23
18	72	69	51	5	15	4	6	33	48
41	93	83	93	99	4	12	3	90	66
14	62	76	42	94	75	57	28	53	4
83	76	75	1	54	70	25	29	23	65
58	84	3	92	37	22	48	20	25	86
21	33	7	56	50	95	13	37	53	36
69	45	66	30	34	26	99	4	67	20
37	17	42	44	79	10	82	96	80	86
92	28	60	96	46	51	72	72	93	77
35	44	53	13	71	84	89	50	25	17
9	5	69	42	89	79	45	59	88	1
21	6	50	88	51	84	88	92	62	64
84	76	16	43	92	74	71	7	22	60
2	6	45	26	29	98	47	38	3	66
6	52	69	81	100	4	32	99	32	6
5	73	65	99	25	13	76	52	48	84
36	32	33	71	18	48	53	43	76	21
54	41	57	49	96	39	25	37	98	97
99	83	72	79	99	32	41	68	68	88
69	76	11	52	4	91	37	44	6	36
1	45	31	28	85	75	4	42	48	97
34	13	66	53	5	76	72	12	12	56
92	12	32	11	53	74	1	4	51	25
7	38	58	33	38	69	66	31	83	60
14	87	22	15	10	29	45	45	84	11
38	25	54	21	26	36	12	76	77	40
57	19	70	67	21	66	86	26	91	61
66	50	100	58	71	26	78	86	54	99
35	63	22	32	72	1	26	75	51	79
4	21	52	69	58	66	36	13	20	94
28	11	42	15	78	78	99	74	60	84
17	33	85	52	34	14	15	82	100	11
96	20	11	34	26	44	36	62	13	39
18	57	7	72	28	34	72	41	77	50
98	17	68	88	47	50	21	61	82	74
5	71	57	24	52	9	26	75	51	47
50	19	1	75	36	60	75	83	45	51
45	15	87	13	56	80	69	57	31	96
63	29	75	73	36	31	31	21	59	100
95	63	31	65	33	13	8	82	0	8
79	28	85	96	35	13	31	60	83	38
48	44	74	49	81	89	2	70	8	54

Appendix C: Some Sources of Air Pollutants

Pollutants   **Sources**

Group I -- Sources predominantly outdoor:

Pollutants	Sources
Sulfur oxides (gases, particles)	Fuel combustion, smelters
Ozone	Photochemical reactions
Pollens	Trees, grass, weeds, plants
Lead, manganese	Automobiles
Calcium, chlorine, silicon, cadmium	Suspension of soils or industrial emission
Organic substances	Petrochemical solvents, natural sources, vaporization of unburned fuel

Group II -- Sources both indoor and outdoor:

Pollutants	Sources
Nitric oxide, nitrogen dioxide	Fuel-burning
Carbon monoxide	Fuel-burning
Carbon dioxide	Metabolic activity, combustion
Particles	Resuspension, condensation of vapors and combustion
Water vapor	Biologic activity, combustion, evaporation
Organic substances	Volatilization, combustion, paint, metabolic action, pesticides, insecticides, fungicide
Spores	Fungi, molds

Group III -- Sources predominantly indoor:

Pollutants	Sources
Radon	Building construction materials (concrete, stone), water
Formaldehyde	Particleboard, insulation, furnishings, tobacco smoke
Asbestos, mineral, and synthetic fibers	Fire-retardant, acoustic, thermal, or electric insulation
Organic substances	Adhesives, solvents, cooking, cosmetics, solvents
Ammonia	Metabolic activity, cleaning products
Polycyclic hydrocarbons, arsenic, nicotine, acrolein, etc.	Tobacco smoke
Mercury	Fungicides, in paints, spills in dental-care facilities or laboratories, thermometer breakage
Aerosols	Consumer Products
Viable organisms	Infections
Allergens	House dust, animal dander

Appendix D: Notes to the Instructor

Ch. 2   Community Structure

A. For information on sampling different habitats, consult Myers, 1980.

B. For information on classification of habitats, consult Southwood, 1954.

C. To obtain information on the major plant species inhabiting the community types, it will be helpful to:

1. obtain field guides to local vegetation, diagnostic keys, and/or actual plant specimens.
2. prepare a checklist of the most common species, and an abbreviated key to take into the field.

Ch. 3   Habitat and Niche

A. Have your communities located before class begins.

B. Be familiar with the major tree species and be able to identify them.

Ch. 4   Ecological Competition

A. Locate the wooded community before class begins.

B. Be familiar with the major tree species and be able to identify them.

C. Before class, determine the maximum distance allowed between trees in a pair; this will depend on the tree species involved.

D. Be careful when out in the field to identify and stay away from poisonous plants (i.e., poison ivy, poison oak).

E. Be familiar with using a random numbers chart and be comfortable with manipulating the calculations for rescaling trees involved in interspecific competition.

F. If you are unfamiliar with using the random numbers chart, you may use an alternative way of finding tree-pair combinations.

Ch. 5   Community Succession

A. Locate the wooded community before class begins.

B. Be familiar with the major vegetative species and be able to identify them.

Ch. 6   Population Dynamics

A. Start fruit fly colonies 2 weeks before this lab to ensure sufficient time for flies to reproduce.

B. Be sure you are familiar with caring for and sexing fruit flies. The biology department, or genetics instructor should be able to help you. In fact, most universities have colonies of fruit flies always in supply.

C. After the lab, return the flies to the biology department, or dispose of them according to university rules.

D. Be able to understand, to work with, and manipulate the population formula.

Ch. 7   Human Population

A. Be able to understand and work with the population formula as applied to humans.

Ch. 8   Water Pollution

A. Be familiar with handling dangerous chemicals and with titration techniques.

Ch. 9   Air Pollution

A Be familiar with working with the Hi-volume air sampler, as, for example, the "Staplex TFIA" unit.

Ch. 10   Soil Management

A. Locate the study area before class begins.

B. Be familiar with manipulating the soil probe and samples.

C. Be familiar with conducting the chemical analysis of soil.

Ch. 11   Renewable Energy

A. Since the intensity of the sun varies from place to place, and from season to season, perform the procedure before class to get the best lab results for your area.

B. Put the water-heating containers in the sun the first thing in the morning, so they will have time to heat up before class.

Ch. 12   Our Finite Resources: The Current Affair

A. Be familiar with your library and its types and sources of information.

B. Encourage students to be original, and to spend enough time on this project to do a good job.

Ch. 13   Environmental Survey

A. Encourage students to survey a topic of relevance in your community.

B. Be sure to include all students in developing the survey.

Ch. 14   Environmental Awareness and Lifestyle

A. This chapter encourages and stimulates community awareness, and individual involvement. Be sure to include each student in the class discussion.

B. Make arrangements before class to visit one or more community facilities. Many places will be happy to give tours if given enough prior notice.